In Search of Solutions

Compiled by Karen Davies
Edited by John Johnston and Neville Reed
Designed by Imogen Bertin
Original design concept by Gordon McSweeney and Nigel Morrison

Published by the Education Division, The Royal Society of Chemistry

For further information on other educational activities undertaken by the Royal
Society of Chemistry write to:

The Education Department
The Royal Society of Chemistry
Burlington House
Piccadilly
London W1V OBN

British Library Cataloguing in Publication Data
"In search of solutions" : some ideas for chemical egg races and other
problem-solving activities in chemistry.
1. Chemistry
I. Royal Society of Chemistry
540

ISBN 1–870343–15–8

THE ROYAL
SOCIETY OF
CHEMISTRY

Contents

THE ROYAL
SOCIETY OF
CHEMISTRY

Foreword

This collection of problem solving activities is an example of curriculum materials written by teachers for teachers. The project arose from a challenge by one of the Society's members for teachers to produce 'chemical egg races' to complement those 'egg races' requiring solutions based on the principles of physics and engineering. Not surprisingly, chemistry teachers throughout the UK rose to the challenge and this book represents the culmination of the efforts of many. As President of the Royal Society of Chemistry, I would like to pay tribute to those teachers who have sent in ideas, and the willing volunteers who trialled the activities in schools. The level of enthusiasm and innovation shown by teachers bodes well for the future of chemistry teaching in this country.

Sir Rex Richards CChem FRSC FBA FRS
President, The Royal Society of Chemistry
December 1990

THE ROYAL
SOCIETY OF
CHEMISTRY

In search of solutions

**Some ideas for chemical egg races and other
practical problem solving activities in chemistry**

by Peter Borrows

Introduction

History of this Book

The idea of "egg-racing" started in the mid-1970s, with a challenge set by BBC Television: how far can you move one (intact!) egg with the energy stored in one rubber band ? Astonishingly far, was the answer, using thoughtful engineering design, strong lightweight materials, low friction bearings and other technological wonders. Many other problem solving activities followed: construct a bridge between two tables 1 m apart, capable of supporting as great a load as possible, using only two sheets of newspaper and 50 cm of adhesive tape; or what is the tallest tower, capable of supporting the weight of an Oxo cube, that can be constructed from dressmakers' pins and drinking straws?

In the early 1980s "egg-racing" took off. Not only did the BBC competition continue, but clones sprang up around the UK SATROs (Science And Technology Regional Organisations) ran many of them, and they even featured in GCSE training sessions. Such events appeared frequently in the programmes of BAYS (British Association of Young Scientists) branches and, as a result, the BAAS (British Association for the Advancement of Science) published two collections[1,2] of ideas for such problem solving activities. However, a perusal of the 200 or so ideas listed in these books showed none requiring the use of any chemical principles. Concerned that physicists seemed to be having all the fun, and worried that chemistry was missing out, I wrote an article in Chemistry in Britain[3]. As a result of this, Malcolm Frazer, then Professor of Chemical Education at the University of East Anglia, persuaded the Royal Society of Chemistry to sponsor a working party to try and produce some CHEMICAL egg races, to put the FUN back into chemistry[5,6].

When the Working Party started to meet it quickly became apparent why chemical egg races were scarce – it is very difficult to devise appropriate tasks. It is curious that it has proved so difficult to do this in chemistry, that supremely technological activity[4]. Nevertheless, thanks to the contributions of many chemists, a bank of ideas was gradually built up and some were tested – through local Chemistry Teachers' Centres, ASE (Association for Science Education) meetings, BAYS Days *etc.* Progress was slow, but in 1989 Dr Neville Reed, newly appointed Schools Liaison Officer for the Society, gained funding from the Society for a School Teacher Fellow. One of the main tasks of that Fellow was to bring the collection of ideas to publication. Karen Davies started in September 1989... and this book is the product.

It may be problem solving, but is it Chemistry?

The original intention was to publish a collection of ideas for chemical egg races, *ie* for enjoyable, competitive, practical problem solving activities which required some application of chemical principles. Many of the previously published ideas for egg races involved constructional projects of some sort: towers, bridges, *etc.* There is no very obvious chemical equivalent. However, another common theme is the use of a rubber band - to drive a vehicle, to raise a load, and so on. This book introduces the idea of the 'chemical rubber band'. Anything you can do with a rubber band, you can do with one teaspoon of sodium hydrogencarbonate (bicarbonate) and three teaspoons of anhydrous 2-hydroxypropane-1,2,3-tricarboxylic acid (citric acid): a roughly stoichiometric mixture. You can drive a vehicle as fast as possible, or as far as possible - although on the whole a boat is preferable to wheeled or flying vehicles, because the mess is then confined to a tank of water. But is this chemistry ? The amount of chemical understanding required is minimal, although some knowledge of handling gases is helpful. At least a chemical reaction is involved, and in any case most of the published egg races require very little explicit knowledge of physics.

However, there are only a limited number of things you can do with a (chemical) rubber band. Once you get beyond that stage you are into 'real' chemistry, which poses problems about how much you can reasonably assume that any particular target group actually knows – but it's always less than you would think. There are ways around this. For example, you can "sell" contestants information by deducting points from their final score. Or you can warn them in advance that they may need to have met the idea of, say, gradually adding measured amounts of an acid to an alkali to neutralise it (as in "Oranges and Lemons"), without really giving too much of the game away. Alternatively, you can allow "open book" competitions (as in "A Salt and Battery").

Competitions versus Investigations

Some activities produce a clear winner, *eg* the most accurate chemical clock, the largest current. Others do not. Thus as ideas were collected and tested, it became apparent that some desirable activities did not lend themselves to a competitive situation. They were enjoyable, they involved practical problem solving with a chemical flavour, and we felt therefore that they should be included. The collection gradually increased to include a range of non-competitive problem solving activities.

The idea of all this is to attract young people into chemistry by showing them that it can be fun. Some of the activities in this book are ones which teachers have been doing for years. In recent years, problem solving has become more prevalent as a day to day laboratory activity. So where does egg-racing end, and straightforward practical chemical

THE ROYAL
SOCIETY OF
CHEMISTRY

problem solving begin ? Perhaps it doesn't matter too much. The tower made of pins and straws started out as an egg race, but later became incorporated into a training activity for teachers, and is now doubtless forming a part of the practical assessment of thousands of pupils. It is also worth remembering that one of the aims of the GCSE National Criteria for Chemistry[7] is "...to stimulate students and create and sustain their interest in, and enjoyment of, chemistry..."

Of course, with the arrival of the National Curriculum in parts of the UK, the situation changes yet again. There is now a specific requirement[8] for collaborative and investigative work in Attainment Target 1, and some of the problem solving activities in this book would help develop such skills. At the higher levels of Science Attainment Target 1 there is a clear requirement for extended investigations; a holistic approach rather than individual skills. Chemistry teachers in the past have been very good at dividing their subject into discrete chunks ("Today, we are going to look at the effect of acids on metals, tomorrow at the effect of acids on hydroxides, and the day after at the effect of acids on carbonates..."), but real life problems don't come neatly quantised in that way. There will be a need, for Attainment Target 1, to develop some interesting and realistic explorations, and it may be that some of the ideas in this book could provide a starting point. We hope that teachers will want to incorporate some of the tasks from this book into their day to day teaching, and if this happens, we hope they will write to the Royal Society of Chemistry and tell us about it, and about any new problems they devise. That way, before too long, we may be able to bring out a second volume of ideas, contributed by our readers and, perhaps, by their students.

There is, however, no reason why pupils should have all the fun. Adults enjoy these too, as we have found out when trying them out on teachers. Some schools may want to consider using these activities on open days for parents or governors, to give them an idea of what modern science teaching is about. The Royal Society of Chemistry is very concerned at promoting the public understanding of science. What better way of doing this than through a practical problem solving event?

Have fun!

> Dr Peter Borrows
> Chairman, 1987-90
> The Royal Society of Chemistry's Great
> Chemical Egg Race Working Party

References

1   Ideas for Egg Races, British Association for the Advancement of Science, 1983.

2   More Ideas for Egg Races, British Association for the Advancement of Science, 1985.

3   P. Borrows, Chem. Br., 1985, **21**, 635.

4   P. Borrows, Educ. Chem., 1988, **25**, 164.

THE ROYAL
SOCIETY OF
CHEMISTRY

5   Educ. Chem., 1986, **23**, 100.

6   Educ. Chem., 1987, **24**, 35.

7   GCSE: The National Criteria, Chemistry. London: HMSO, 1985.

8   Science in the National Curriculum. Department of Education and Science and the Welsh Office. London: HMSO, 1989.

Acknowledgements

The membership of the Working Party fluctuated somewhat over its lifetime. Some members' contributions mainly took the form of ideas, and comments on other people's ideas. Some were co-opted onto the Working Party as a result of having tried a chemical egg race at an ASE or similar meeting. Such meetings were also a fruitful source of new ideas, and of volunteers willing to test them out in their own schools. Originators of ideas for chemical egg races are acknowledged on the relevant page, where we have been able to identify them, although sometimes they may have been modified (almost out of recognition) as a result of trialling. To all who contributed in any way – our most sincere thanks. Please accept our apologies if your contribution has not been fully acknowledged – we will try to get it right next time.

The following served, at some stage, on the Working Party :

Dr Peter Borrows
Dr John Crellin
Ms Karen Davies
Mr Justin Dillon
Dr Malcolm Frazer
Mr Colin Johnson
Professor Richard Kempa
Mr Dick Langstaff
Mr John Lawrence
Mr Ron Lewin
Mrs Sue Lindley
Dr Sue Pringle
Dr Neville Reed
Mr Jack Roberts
Professor Jeff Thompson

Particular thanks are due to Karen Davies, without whose work this publication would not have come to fruition.

Thanks are also due to Esso UK plc for helping to fund the production of this publication.

THE ROYAL
SOCIETY OF
CHEMISTRY

Acknowledgements

The Society would like to thank the staff and students of the following schools/colleges for their help in trialling the experiments in this book:

Archbishop's Church of England School, Canterbury, Kent
Bells Brae Primary School, Lerwick, Shetland
Carre's Grammar School, Sleaford, Lincolnshire
Chipping Sodbury School, Chipping Sodbury, Avon
Farnham College, Farnham, Surrey
Gladys Buxton School, Dronfield, Derbyshire
Godalming College, Godalming, Surrey
The Godolphin School, Salisbury, Wiltshire
Hall Cross Comprehensive School, Doncaster, South Yorkshire
Heathfield School, Monkton Heathfield, Somerset
Jack Hunt School, Peterborough, Cambridgeshire
King's College,Centre for Educational Studies, London
King Edward's School, Birmingham, West Midlands
The King Edward VI School, Morpeth, Northumberland
Leasowes High School, Halesowen, West Midlands
Ysgol Uwchradd Llanidloes, Llanidloes, Powys
Ilford County High School for Boys, Essex
Monks Park School, Bristol, Avon
Netherthorpe School, Staveley, Derbyshire
Philips High School, Bury, Greater Manchester
Prestwich High School, Prestwich, Greater Manchester
Queen Mary's College, Basingstoke, Hampshire
Richard Aldworth School, Basingstoke, Hampshire
St Mary's School, Calne, Wiltshire
St Mary Redcliffe and Temple School, Bristol, Avon
Stamford High School, Stamford, Lincolnshire
Stanborough Secondary School, Welwyn Garden City, Herts
International School, Vienna
Tavistock College, Tavistock, Devon
The University of Leeds, Centre for Studies in Science & Mathematics
   Education, Leeds
Wilson's School, Wallington, Surrey

The Society also thank the many individuals who have given their time and effort to the project, in particular:

Nan Davies, Vivienne Davies, Martin Goodall, Ian Harrison, Bill Pritchard, Ray Vincent and Martin Wesley.

THE ROYAL
SOCIETY OF
CHEMISTRY

Practical problem solving – 'Advice for Teachers'

by Karen Davies

"If we do not experiment with concepts they will remain remote theory, or as much use as hieroglyphs to the average man".
(Charles Handy,
"Understanding Organisations".
London: Penguin Books Ltd., 1985)

The main aim in writing this book was to provide teachers with a useful resource of chemically-based problem solving activities, egg race style experiments and further 'ideas' for use in the classroom – all of which highlight the 'fun' of chemistry. In choosing experiments for inclusion the Society has taken a broad based interpretation of what constitutes 'chemistry' and has tried to include a few experiments suitable for children of primary school age.

The experiments are grouped together in "topic categories", *eg* acids and bases, electrochemistry, separating substances *etc*, so that they can be easily accessed by users. They can be used not only to enhance a topic taught in 'lesson time', but also as an end of term activity, in science clubs, or on open days (or open evenings) where parents could take part in a 'problem solving' activity, thus encouraging, in a wider context, the public understanding of science. The Society also hopes that any groups organising large science events for schools will find this book useful.

Most of the material is suitable for further development, and only the more obvious curriculum links have been highlighted. The book is not meant to be prescriptive and there is scope for developing the experiments to suit individual needs and situations.
In my own teaching I found problem solving to be a very rewarding, but demanding, experience. It can be used to:

- ☛ enhance the understanding of the concept being studied. In this respect it acts very much as a reinforcing tool since it requires students to apply and, therefore, understand their acquired knowledge;
- ☛ gauge students' understanding of their work, and can thus be incorporated into student assessment;
- ☛ develop students' social and communication skills since it requires them to work in project teams;
- ☛ build CONFIDENCE and aid MOTIVATION.

Furthermore problem solving encourages students to use the scientific principles and concepts that they have been taught and equips them with skills that can be applied in everyday life, to a wide variety of situations.

THE ROYAL
SOCIETY OF
CHEMISTRY

However, problem solving does not provide a single, magical, answer to the problem of stimulating and maintaining students' interest in chemistry: one teaching method does not meet the needs of every child. There are many ways to learn, and although problem solving places many demands on students (as well as on teachers and technician support) it is not advocated that problem solving should be used in every lesson. Instead problem solving should be seen as one of several complementary techniques available for use by the science teacher.

As teachers we also need to look carefully at the ways in which we get students to report on their work. There are many options open to us: poetry, drawing, letters, plays, songs, interviewing; so why do we often present them with only one option? In this respect many of the activities included in this book lend themselves to ways of presenting results other than the straightforward 'what we did' approach.

All of the experiments have been trialled in schools around the UK and the comments within quotations are feedback from those schools.

The "Teachers' Notes" provide an indication of some of the approaches used by students and teachers when tackling the different problems. (However, it is wise to keep an open mind and to encourage lateral thinking.) When using problem solving in the classroom try to ensure that students don't change too many variables at once, but rather adopt a step-by-step approach. Safety should be paramount and it is essential that the importance of safe working practices is stressed. Students should also be encouraged to consider safety aspects in the planning of their experiments and all students should have their ideas checked by the teacher at the planning stage, before being allowed to put their plans into action.

Equipment that all participants need can be put out at each work place. Generally, however, choosing suitable equipment or chemicals is part of the problem to be solved. Therefore have a range of items, including distractors, on tables/benches at the side. These should be arranged in a logical way, *eg* separating chemicals from equipment. Experience also suggests that for experiments requiring junk items it is better to display these separate from non-junk items, such as laboratory equipment.

Problem solving activities are not meant to be carried out by individuals, so students should be organised into groups the size of which will vary depending on the situation. Two may not be a suitable size: it is not fair on the student who has to partner a deadweight! However, if the size of the group gets too big, too many people sit around, without having 'hands-on' experience. Through discussion, brainstorming and the sharing of ideas and tasks, students will be ready to take up the challenge provided by the experiment – as a result they will get a feel for what it is like to work in a scientific team.

THE ROYAL
SOCIETY OF
CHEMISTRY

The problem solving process

1   Identify the problem

2   Identify possible solutions

3   Choose best solution

4   Design experiment ◄──────────── 8   Re-evaluate and
                                            modify
                                            experiment

5   Safety check (by teacher)   ▲
                                                  │
6   Carry out experiment ──────────────► 7b   Unsuccessful
│   solution
▼
7a   Successful Solution

THE ROYAL
SOCIETY OF
CHEMISTRY

Safety

Safety can cause difficulties in open-ended problem solving activities. It is in the nature of a problem that you cannot always anticipate what a participant will do. Indeed, some of the best solutions during trials of these problems were the unexpected ones. But children can do quite bizarre things. Most teachers will have had the experience of asking a class to measure the temperature of water being heated, only to see one child stick the thermometer straight into the Bunsen flame.

Teachers need to be particularly vigilant during practical problem solving activities, especially when chemicals are involved. A higher degree of supervision is needed than in activities which have more closed outcomes. Students must be encouraged to take a responsible attitude towards safety, both their own and that of others. A statement to this effect could appear prominently in the instructions for the problem. In planning an activity students should always include safety as a factor to be considered. Plans should be checked by the teacher before implementing them (unless chemicals and equipment are so constrained as to make that unnecessary). Remember, however, you are not checking whether it will work, but whether it is safe. Always insist upon eye protection. Even if a student claims to be still thinking, the one on the opposite side of the room may not be! Judges must set a good example themselves. During a problem solving event, and especially if artefacts are being tested in some way at the end, there can be a high level of excitement. Do not allow things to get out of hand, or competitors to put themselves in positions of potential danger.

Under the COSHH Regulations (Control of Substances Hazardous to Health) there is an obligation upon employers to carry out a risk assessment when hazardous substances are used. Most LEAs (Local Education Authorities) and many independent schools have followed the recommendations of the Health and Safety Commission's Education Service Advisory Committee, given in 'COSHH: Guidance for Schools' and adopted certain publications as the basis for their General Risk Assessments. Those most commonly used are:

Topics in Safety (ASE, 2nd edn, 1988)
Hazcards (CLEAPSS, 1989 edn)
Safeguards in the School Laboratory (ASE, 9th edn, 1988)

In general, we have tried to stay in line with these standards. The more unusual the activity, the less hazardous are the chemicals suggested for use. Nevertheless, teachers should always carry out a risk assessment in accordance with the procedures of their employer, and they may decide that some suggested activities are inappropriate in their situation.

THE ROYAL
SOCIETY OF
CHEMISTRY

How to judge chemical egg races and other problem solving activities

by Ron Lewin

When judging chemical egg races and other problem solving activities, the aim of the activity should be carefully considered. Originally such activities were developed to raise an awareness of the challenge and excitement of science and technology. However, with the implementation of the National Curriculum Science and Technology documents the activities can usefully enhance the attitudes, qualities and skills needed by scientists of all ages.

Judging projects involves both objective and subjective decisions. It is important to ensure that the project is carried out in a purposeful way; that both the problem and the relevant scientific and technological principles are understood; that a range of alternatives are considered; and that the experimental programme reflects the information collected. It is also necessary to consider how the experimental data was collected, displayed and finally what conclusions were drawn.

Of equal importance in judging projects are subjective factors such as ingenuity, curiosity, novelty, enthusiasm, commitment, perseverance, team work, practical ability and an aesthetic sense. Remember that egg race projects/problem solving activities have been developed to encourage young people to take part in scientific pursuits in a warm and encouraging way. We should expect scientific rigour, relevant to age and ability, but at the same time the activity should be enjoyable.

If the above points are borne in mind, egg races/problem solving activities can enrich and complement more formal school science. Not all the factors will need to be included for each activity and teachers will need to choose those relevant to the particular occasion. To avoid misunderstanding, it is worth explaining to participants the basis of the assessment and how marks will be awarded. The following is a list of criteria for which marks could be awarded:

- Understanding the problem
- Use of scientific method
- Collection of information
- Consideration of alternative procedures
- Experimental design
- Practical work
- Recording information
- Interpretation/critical assessment of results
- Success in solving the problem
- Suggestions for further work
- Co-operation, team work (if relevant)
- Originality, novelty and ingenuity

☛ Curiosity and inquisitiveness
☛ Aesthetic sense
☛ Perseverance
☛ .. .. ..

Teachers must decide which to use, how they should be weighted and whether any other factors need to be taken into account.

THE ROYAL
SOCIETY OF
CHEMISTRY

Links with the curriculum

by Neville Reed

The majority of activities in this book can be used, or adapted for use, to deliver parts of the school curriculum. By their very nature all of the activities are open to modification to suit pupil and teacher needs; in fact it is hoped that the activities will act as a starting point for better ideas and experiments.

Within schools throughout the UK the school science curriculum has, over recent years, placed increased emphasis on problem solving and team work. The activities collected together in this book rely on both problem solving and team work for their solution, hence most teachers should find this book a valuable one to dip into. Although the collection is deliberately biased towards problems using chemistry as a solution, other disciplines such as technology will find ideas of interest to them.

The activities have been grouped together into chemical topics for convenience and ease of reference. Given the statutory requirements of the science curricula in England, Wales and N. Ireland, some cross references to attainment targets for each science curriculum have been attempted. In no way should this cross referencing be seen as comprehensive, rather it should be treated as no more than helpful suggestions. It is anticipated that teachers will modify the activities to suit their own particular requirements both in science and other subjects.

In its origin, the project developed from the idea that chemical egg races were fun and it is hoped that in whatever context these activities are used they will stimulate interest and enthusiasm, and that participants will enjoy finding a solution to the problem. The Royal Society of Chemistry hopes this goal will be met.

Links with the national curriculum in science for England and Wales

Most of the activities will have an application supporting the Programme of Study for Attainment Target 1, and below are suggestions where the activities may be useful in delivering other parts of the science curriculum.

Attainment targets / Egg race	2	3	4	5	6	7	8	9	10	11	12	13	14	15	16	17
1					5b											
2					5b											
3					5a	6a										
4					5b,6a											

continued opposite

THE ROYAL
SOCIETY OF
CHEMISTRY

Attainment targets / Egg race	2	3	4	5	6	7	8	9	10	11	12	13	14	15	16	17
5					3a, 6a											
6				6ab	5b	6a	8d, 9a									
7					5bc											
8																
9					5b	4b										
10					6d	6b				7b						
11					6d	6b				7b						
12					5c	6b				3ab, 7b						
13						4a, 6ab, 10b				7b						
14						6c						5a, 7a				
15						6c						4bd				
16						6c										
17						6c						4bd, 5a				
18						6c						5a, 7a				
19					4c				3b							
20									6b, 7b							
21									4a, 6b, 7b			4c				
22									6b, 7b							
23						7a										
24					6ae, 8a											
25						7a, 8a										
26					6ae, 8a											
27				2a												
28						4a										
29					4a, 5b, 6ac	4a										
30						7a, 8a										
31						7a, 8a										
32						7a, 8a										
33					8ab	7a				5a						
34					6d, 7c, 9a											
35					2ab, 4a, 5c				3b							
36		4b			5c											
37					5c											
38				6ab	4d, 5c		6b, 7a	5e								
39					5c											
40					4d, 5c, 6b											

continued overleaf

THE ROYAL
SOCIETY OF
CHEMISTRY

Attainment targets / Egg race	2	3	4	5	6	7	8	9	10	11	12	13	14	15	16	17
41					5c											
42																
43					4ab											
44									3a, 4a							
45																
46					7ab											
47					4abce, 5abc											
48		6b, 7a														
49					4a, 6a											
50																

☞ numbers represent levels of attainment and letters statement of attainment.

Links with the national curriculum in science for Northern Ireland

Most of the activities will have an application supporting the Programme of Study for Attainment Target 1, and below are suggestions where the activities may be useful in delivering other parts of the science curriculum.

Attainment targets / Egg race	2	3	4	5	6	7	8	9	10	11	12	13	14	15	16
1								5d							
2								5d							
3								5d	6b						
4								5d, 6b							
5								3a, 5d							
6							6ab	5d	6b	8b, 9b					
7								5d, 6e							
8															
9								5d	5a						
10								6c	6c						
11								6c	6c						
12								6e	6c			4ab			
13									6abc, 8d						
14			5a, 7b						6e						
15			4be						6e						
16									6e						
17			4be, 5a						6e						

continued opposite

THE ROYAL
SOCIETY OF
CHEMISTRY

Egg race \ Attainment targets	2	3	4	5	6	7	8	9	10	11	12	13	14	15	16
18			5a, 7b						6e						
19											3b				
20											6c, 7b				
21											4a, 6c, 7b				
22											6c, 7b				
23									7a						
24									6ad, 8c						
25										7a, 8a					
26									6ad, 8c						
27								3a							
28									6a						
29							4a, 5d, 6a								
30								7a, 8a							
31									7a, 8a						
32									7a, 8a						
33								8bc	7a			5a			
34								6c, 7d, 9b							
35								2bc, 4a, 6e							
36					4b			6e							
37								6e							
38							6ab	6e		4b, 6b, 7a					
39								6e							
40								6be		4b					
41								6e							
42															
43								4ab							
44										3b, 4a					
45															
46								7bc							
47								4abcd, 5ad, 6e							
48					6b, 7a										
49								4a, 6a							
50															

☛ numbers represent levels of attainment and letters statement of attainment.

THE ROYAL
SOCIETY OF
CHEMISTRY

"In search of solutions"

Below is a list of egg race titles. Each egg race has a student sheet (labelled S) and a teacher sheet (labelled T). Numbers in italics represent egg races which are applicable to more than one heading.

Egg race	Title
Acids and alkalis	
1	Making your own indicator
2	Colour creation
3	What is the best indigestion cure?
4	Technician in trouble! Which solutions are which?
5	Indicator puzzle
6	The duck pond problem
7	Funny felt pens
8	Oranges and lemons *(9, 46)*
Colour chemistry	
9	Natural dyeing
	(2, 7, 39)
Density	
	(49)
Electrochemistry	
10	Kitchen potential
11	Kitchen currents
12	Extracting lead from "lead ore" and bridge that gap!
13	A salt and battery
Energy	
14	The 'cooking an egg using the least energy' challenge
15	Heating power of a candle
16	Cooking an egg by a chemical reaction
17	Which fuel is better?
18	The candle cooker
Exothermic reactions	
	(16)
Gases – power	
19	The ups and downs of chemistry
20	Lift Oxo cubes to dizzy heights!
21	Building a chemically-powered boat
22	The heavy lift cup challenge
	(46)

THE ROYAL
SOCIETY OF
CHEMISTRY

THE ROYAL
SOCIETY OF
CHEMISTRY

Junk list

Many of the chemical egg races require the use of 'junk'. The following is a list of the type of item envisaged:

plastic lemonade bottles
'squeezy' bottles (washing up liquid containers)
empty beer/soft drink cans (dry)
coffee tins/syrup tins
coffee jars/jam jars
yoghurt pots/margarine tubs
shoe boxes/cereal packets
cardboard tubes from toilet rolls/kitchen towels
blocks of expanded polystyrene packing
polystyrene meat trays/egg boxes
disposable foil trays (oven ready)
lollipop sticks
wood off-cuts/cotton reels
used tights (empty!)

NON-JUNK ITEMS often used alongside junk:

sticky tape
glue &/or glue gun
blu-tack/plasticine
string
rubber bands
paper clips/split fasteners
pegs
wire
pins
aluminium kitchen foil
cling film
balloons
plastic bags
drinking straws
plastic tubing
assorted bungs & corks
plastic syringes
plastic gloves
paper towels
stapler
ruler
simple tools: tin snips, saw, bradawl, file, stanley knife *etc.*

THE ROYAL
SOCIETY OF
CHEMISTRY

We have run out of Universal indicator stock; but we have a selection of dyes.

☞ **Your task**

Select from these dyes and make a suitable indicator.

The indicator that you make must be able to distinguish between:-

strong acid
weak acid
neutral
weak alkali
strong alkali

▲ Use only small quantities of solutions.

Based on a suggestion by R.F. Kempa.

THE ROYAL
SOCIETY OF
CHEMISTRY

Age	12–16 years.
Time	70 minutes.
Group size	2–3.
Equipment & materials	Eye protection.

Per group
Test tubes and racks or spotting tiles, plastic droppers, small measuring cylinder, selection of beakers, glass stirring rod, scissors, labels.
Labelled solutions:
strong acid (sulphuric acid) .... – 50 cm^3
weak acid (ethanoic acid) .... – 50 cm^3
distilled water or tap water (**NB** check that water is neutral)
weak alkali (sodium carbonate) .... – 50 cm^3
strong alkali (sodium hydroxide) .... – 50 cm^3
dye solutions A to H:

Must include

A= methyl red .... – 5 cm^3	E= litmus .... – 5 cm^3
B= phenolphthalein .... – 5 cm^3	F= a red food dye .... – 5 cm^3
C= thymol blue .... – 5 cm^3	G= a blue food dye .... – 5 cm^3
D= bromo-thymol blue.... – 5 cm^3	H= a green food dye ...... – 5 cm^3

☞ **NB** Emphasise to students that only small quantities of solutions are to be used.

Safety notes	See page 11.
Curriculum links	Indicators. Acids and alkalis.
Possible approaches	An answer grid could be given to less able students so that they start by testing all the solutions with all the indicators. Although the food dyes are not essential to the indicator made, they are a useful distraction. Spotting tiles would be easier to use than test tubes.
Evaluation of solution	Whether an indicator mixture was found that does the job.

Your notes

....

....

....

....

....

THE ROYAL
SOCIETY OF
CHEMISTRY

THE ROYAL
SOCIETY OF
CHEMISTRY

☛ Your task

Starting from Universal indicator solution, water, hydrochloric acid solution and sodium hydroxide solution only:

1   Produce six different coloured solutions (*ie* red, orange, yellow, green, blue and violet).

2   When you have completed Part 1 (or after 20 minutes), produce a reliable "recipe" for creating one colour decided by your teacher.

Based on a suggestion by A. Honeyman.

THE ROYAL
SOCIETY OF
CHEMISTRY

Age	13–16 years.
Time	60 minutes.
Group size	2–3.
Equipment & materials	Eye protection.

Per group
Rack of test tubes, glass droppers, beakers (100 cm^3), measuring cylinders (l0 cm^3).
Universal indicator solution .... – 5 cm^3,
distilled water or tap water (**NB** check that the water is neutral),
Hydrochloric acid (0.1 mol dm^{-3}) .... – 50 cm^3,
Sodium hydroxide (0.1 mol dm^{-3}) .... – 50 cm^3.

Safety notes	See page 11.
Curriculum links	Indicators. Acids and alkalis.
Possible approaches	Useful as an end of term activity. Enables students to gain an appreciation of the nature of indicators. (Demonstration of the effects of acid and alkali on Universal indicator may be needed for younger groups.) Students tend to get bored if they are not meeting with success, especially if they have no plan and are mixing at random. Students may need encouragement to tackle the experiment in a systematic way. They may also need to be helped to see that further dilution of the acid and alkali will result in less abrupt colour changes (alternatively students could be given more dilute acid and alkali). Some students found it very difficult to produce the 6 colours. To avoid the possible difficulties associated with different shades of the same colour, a constant volume should be presented in the tubes for judging (*eg* half full tubes) and also a constant volume of indicator used each time.

Part 2
Each team must show their recipe to the judge. They should then try to make their given colour, following their recipe, under the eyes of the person acting as judge. One teacher commented: "the colour I chose for the recipe was yellow – green is easy and orange is almost impossible!".

Evaluation of solution These are suggestions only:

Part 1
Credit could be awarded for each different colour produced and for the first group to finish.

Part 2
Full credit could be awarded for a "recipe" which "works", *ie* produces the stated colour, first time. A descending order of credit could be awarded for a "recipe" which does not "work" when followed, but which does produce the stated colour after ONE adjustment (*eg* "add 3 more drops of solution X") or for a recipe which does not "work" first time, nor after one adjustment but which produces the stated colour

THE ROYAL
SOCIETY OF
CHEMISTRY

after the second adjustment.
In the event of a tie, the judges could state another colour, and the group which produces it first would be the winner.

Extension work

Give students a weak acid such as ethanoic acid. Students may assume that the same volumes are needed to obtain the colours – this brings over concept of weak versus strong acids.

Your notes

Indigestion is frequently caused by excess acid in the stomach. Many cures work by neutralising this acid.

☞ Your task

You are asked to devise tests to compare various indigestion cures and then to write a report on them for a consumer magazine. Examples of things which you may compare include value for money, effectiveness and speed of action.

NB

1   Bromophenol blue is a convenient indicator for the experiment, turning yellow in acid and blue in alkali.

2   In each test use half of the standard dose of the indigestion cure.

3   When trying out one of your experiments use sodium hydrogencarbonate (since it is the cheapest available cure). Half of the standard dose is equivalent to one teaspoon.

Based on a suggestion by G. Wickens.

THE ROYAL
SOCIETY OF
CHEMISTRY

THE ROYAL
SOCIETY OF
CHEMISTRY

Age	13–16 years. (All abilities – students can take the work to their own level.)
Time	Total time = 140 minutes, but this time can be split into 2 lessons.
Group size	2–3.
Equipment & materials	Eye protection.

General
Burettes, measuring cylinders (100, 250 cm^3), filter funnels, conical flasks, glass droppers, glass stirring rods, pestle and mortars, teaspoons. Balances should be available. Bunsen burners, tripods, gauzes, heatproof mats, clampstands. Thermometers.
As wide a range of indigestion cures as possible, including sodium hydrogencarbonate (sodium bicarbonate).

Per group
Bromophenol blue indicator solution. Bottle of hydrochloric acid (0.5 mol dm^{-3}) – about 500 cm^3 labelled "STOMACH ACID".

Safety notes	See page 11.
Curriculum links	Acids and bases. Neutralisation.
Possible approaches	Best done after burettes have been used. Get students to discuss what they think an antacid is and how it works. What is meant by "best"? ("Best" at what?) Comparing different indigestion cures for effectiveness, speed of action or value for money. Consideration of keeping variables constant for comparison purposes. Exercise enables considerable discussion on advertising, types of remedy.

NB One student suggested that the exercise was done at 37 °C (body temperature). Students could imitate chewing by crushing up the tablet before adding acid.
Prescribed amounts of cure could be used for comparison. Use indicator to show how much acid is needed to neutralise the prescribed amount of cure.

Evaluation of solution	Alternative write-up
An "Indigestion Cure" rap song!
(The 'conclusion' of a rap song composed by Emma Culley, Rebecca Eley and Rebecca McConnell:
"Sodium hydrogencarbonate here
And I want you to know that I'm not so dear,
hey look at all this acid that I can take,
so go out and buy me 'cause I'll
STOP YOUR STOMACH ACHE").
Extension work	Sodium hydroxide will neutralise hydrochloric acid – why couldn't it be sold as an antacid? Also there is the problem of carbonates producing gas. Have students collect advertising claims of various indigestion cures.

THE ROYAL
SOCIETY OF
CHEMISTRY

Your notes

THE ROYAL
SOCIETY OF
CHEMISTRY

The science technician has problems! She has five colourless solutions labelled A, B, C, D and E. She wrote down which solution was which and gave this list to the teacher. BUT ... the teacher has lost the sheet! The technician has to find out quickly the identity of each solution because they are needed for the next class!

She does remember that one solution is water, one is a strong alkali, one is a weak acid, one is a strong acid and one is phenolphthalein. However, the pH meter is broken and the indicator papers are lost! HELP!!!

She needs a quick answer – any group who can solve the problem doesn't get a science homework!

☛ **Your task**

Think how to do the experiment, then ... have a go!

Based on an idea from the APU science question 'indicators'.

THE ROYAL
SOCIETY OF
CHEMISTRY

THE ROYAL
SOCIETY OF
CHEMISTRY

Age	14 years upwards. Higher ability 14–15 year olds (experiment found to be demanding intellectually). A revision exercise for 16–18 year olds. Students need to have met neutralisation before.
Time	40–70 minutes. (Likely to be variable according to ability of students.)
Group size	2–3.
Equipment & materials	Eye protection.

Per group
5 dropping bottles (or test tubes and long glass droppers) labelled A, B, C, D and E. (Each bottle (or test tube) contains a colourless liquid). A test tube rack containing 6 test tubes, glass stirring rod.

Solutions A to E:
A = strong acid (hydrochloric acid)............................20 cm^3
B = water ....20 cm^3
C = strong alkali (sodium hydroxide) ....20 cm^3
D = * weak acid (carbonic acid or tartaric acid).........20 cm^3
E = ** "phenolphthalein indicator + water" mixture . 20 cm^3

* preferably one that doesn't have a recognisable smell. Students can identify vinegar (ethanoic acid) by smell.

** **NB** This needs to be made up on the day its needed as it loses its strength. (Phenolphthalein indicator is made up in alcohol. If you add phenolphthalein/alcohol indicator solution to water, the water goes CLOUDY. You don't want this to happen. You will therefore need to make a "phenolphthalein/alcohol indicator + water" mixture (keep adding water until the solution clears) which:

(i) when added to water, does not make the water cloudy, and
(ii) when added to alkali will go pink.

☞   Recommend keep 250 cm^3 of each solution in stock in case of spillages *etc.*

Safety notes	See page 11.
Curriculum links	Neutralisation. Acids and alkalis.
Possible approaches	At the start of the exercise show students the colour changes for phenolphthalein indicator (colourless in acid and water, pink in alkali). Tell students always to replace the glass droppers in the right bottles (or test tubes) – *ie* point out dangers of contaminating solutions.

A possible sequence of operations:-
Mix each unknown solution with every other unknown solution. The only colour change is pink (E + C = phenolphthalein and alkali, but students won't know which is which). As long as students know the colour changes of phenolphthalein indicator they should get this far. Next, add the other liquids to (E+C):

(E+C), add A – pink solution goes colourless quickly, also test tube feels hot.
(E+C), add B – pink solution remains pink, just becomes more dilute. Therefore B = water.
(E+C), add D – pink solution eventually goes colourless, but more D is needed than A for neutralisation. Therefore D = weak acid, and A = strong acid.

Finally, to find out which is the indicator and which is the alkali:- Mix a small amount of A (strong acid) with a small amount of C (strong alkali) and lots of E (indicator) – solution stays colourless. Then, mix a small amount of A (strong acid) with a small amount of E (indicator) and lots of C (strong alkali) – solution goes pink. This proves that C is the alkali. Alternatively, add a lot of C (strong alkali) to E (indicator) – this gives a weak pink colour. Then, add a lot of E (indicator) to C (strong alkali) – this gives a strong pink colour. This confirms that E is the indicator. (The indicator makes the colour.)

Students should be encouraged to do a flow chart or table.

Your notes

....

....

....

....

....

....

....

....

....

....

....

....

....

THE ROYAL
SOCIETY OF
CHEMISTRY

☞ Your task

Extract colours from given plants and use them to determine
the pH of three given solutions. No indicator paper is allowed!

Based on a suggestion by P. Borrows.

THE ROYAL
SOCIETY OF
CHEMISTRY

Age

14 years upwards. Problem needs to be well structured for lower ability groups (see possible approaches below).

Time

70 minutes.

Group size

3–4.

Equipment & materials

Eye protection.

General

Test tubes and racks (or spotting tiles – if possible, 2 per group), small beakers, pestles and mortars, glass droppers, filter funnels and papers, scissors, labels. Students may need access to hot water (*eg* a kettle, NOT Bunsen burners).
Sand.
Solvents: Ethanol and propanone (acetone).
Samples of coloured plants: Red cabbage, beetroot, geranium petals (strongly recommended because they show two colour changes, at low and high pH), rose petals, violets, delphiniums, lupins, elderberries, blackberries *etc*. Dark-coloured flowers will work best. (Suitable materials could be collected in season and stored in a deep freeze.)
Access to buffer solutions with clearly marked pH values, *eg* pH 2.0, 4.0, 6.0, 8.0, 10.0.
Solutions X, Y, and Z of unknown pH (you could also provide 'buffers' here).

Safety notes

See page 11. Because of the use of flammable solvents, Bunsen burners should not be used. If a hot solution is required, make hot water available from a kettle.

Curriculum links

Indicators. Buffers. Acids and alkalis.

Possible approaches

To extract colours from given plants ("A local florist very helpfully gave us some "newly dead" flowers for free and they were excellent!") and use them to determine the pH values of three given solutions. Initially students could tackle the problem individually to prepare one plant extract, then in groups to solve the X, Y, Z puzzle.

Red cabbage was found to be the best indicator with a sensitive enough response to allow accurate pH estimations (changes through red(cerise) – mauve – blue – turquoise). Other materials (blackberry, beetroot) are interesting to try but very much the same in their response (red up to about pH 6, blue-red above this value). Red rose and carnation were also tried – results seemed to vary, some students finding these more useful than others. **NB** "Especially interesting are the extracts from red cabbage, radish skin, rhubarb skin, and turnip skin, which act as universal indicators" (see "Edible Acid-Base Indicators" Robert C. Mebane and Thomas R. Rybolt. J.Chem.Educ., April 1985, Volume 62, Number 4, p285).

Difficulties with the sequence of events:- Some students will not see the relevance of testing the 'indicator' against known pHs, then using their results to test X, Y and Z. The concepts may need to be explained to

THE ROYAL
SOCIETY OF
CHEMISTRY

them. Also they tend to know universal indicator quite well and assume that all indicators are red in acid and blue in alkali. Teachers could cut the buffers down to only four and label them:- strong acid (pH 3), weak acid (pH 6), weak alkali (pH 8), strong alkali (pH 10). Likely areas for giving guidance:-

how to prepare a plant extract;
how to tackle the testing of the extract systematically;
how to record systematically;
how to interpret results.

Experiments could also be re-stated for lower ability range – maybe as a series of structured steps, to make it easier to understand:-

* * *

"Indicator puzzle"
Can you make an indicator solution and use it to find out the pH of unknown solutions X, Y and Z?
You do have some solutions whose pHs you know – pH 2, pH 4, pH 6, pH 8 & pH 10.

Stage 1
Each person's job is to extract the colour from the plant sample they have been given.

Stage 2
Is your coloured extract an indicator? Try to find out, using the solutions whose pHs we know.

Stage 3
Working in a group, decide which is the most useful indicator to help solve this problem:
What are the pH's of X, Y & Z?
Write down your answers.

* * *

Suggested write-up

Students produce a colourful poster on natural indicators.

Evaluation of solution

Groups who get closest to the pH values of the three solutions are the winners.
Experiment could be used as an assessed practical, *ie* planning & carrying out, group work, and communication (a written report at the end) could all be assessed.

Extension work

Test a variety of household substances with your natural indicator.
Make natural indicator 'papers' by putting the plant pigment onto a piece of blotting paper.

THE ROYAL
SOCIETY OF
CHEMISTRY

During the night a flytipper has dumped a load of waste in the local duck pond which has made it too acidic to support life. The local Forensic Science Laboratory is too stretched at present to carry out a quick analysis and therefore they have approached our school for help.

☞ **Your task**

How can the pond be returned to a habitable condition?

Based on a suggestion by E. Grimble.

THE ROYAL
SOCIETY OF
CHEMISTRY

THE ROYAL
SOCIETY OF
CHEMISTRY

Age	16–18 years.
Time	70 minutes.
Group size	2–3.
Equipment & materials	Eye protection.

General
Titration apparatus. (Surveying equipment for the pond.)

A sample of muddy water with some acid added, sodium hydroxide solution (1.0 mol dm^{-3}). A suitable indicator.

Safety notes	See page 11.
Curriculum links	Acids and alkalis. Neutralisation.
Possible approaches	If you've got a local duck pond the teacher needs to set the scene by asking who walked in past the duck pond this morning? Did anyone notice all the dead fish floating on the surface? If not, you need to talk about the local river (relate to waste dumped in river).

The groups need to determine the equivalent amount of 1.0 mol dm^{-3} sodium hydroxide needed to neutralise the acid. The problem could then be developed to include estimation of the amount of water in the pond and the original concentration of acid dumped in the pond (assuming initial volume dumped is negligible in comparison to large volume of pond!). Finally students need to address the problem of how the pond can be returned to a habitable condition.

Suggested write-up	Students could write a newspaper article about the event ... their own scientific investigation and proposed restoration work ... in the style of a newspaper of their choice.

Extension work

- ☛ Consider biological aspects of pond/river/lake pollution
- ☛ What are the visible signs of pollution?
- ☛ What has happened to lakes in Scandanavia that have become polluted?
- ☛ Consider sources of pollution – *ie* might not be just liquid waste, *eg*
  - sulphuric acid plant
  - aluminium processing plant
  - electroplating company

Students could then carry out other tests to identify the source of the pollution, *eg* sulphate tests. Would it be conclusive – sulphate positive test anyway? Could give background to processes.

THE ROYAL
SOCIETY OF
CHEMISTRY

Your notes

....

....

....

....

....

....

....

....

....

....

....

....

....

....

....

....

....

....

....

....

....

THE ROYAL
SOCIETY OF
CHEMISTRY

You can buy felt pens which show one colour when used normally, but give a second colour when you go over them with a "magic pen".

☞ **Your task**

Find out how these pens work, using paper chromatography.

Based on a suggestion by C.H. Johnson.

THE ROYAL
SOCIETY OF
CHEMISTRY

THE ROYAL
SOCIETY OF
CHEMISTRY

Age	16 years upwards.
Time	70 minutes.
Group size	2–3.
Equipment & materials	Eye protection.

General

A range of apparatus for the commonly used paper chromatographic techniques: Stoppered boiling tube into which filter paper strip can be inserted, beakers (100 cm^3), petri dishes, glass droppers, scissors, paper clips, pencil and ruler (measuring to mm).

1 packet of 'Funny' felt pens – *eg* "Colour Swops" (Platignum) or "Colourscribe" (WH Smith) or "Magic pens" (Asda).
Whatman grade 1 chromatography paper (100 m roll, 30 mm wide) – available from: Griffin & George Limited and Philip Harris. (Alternatively ordinary filter paper may be used.)

Solvents: Water, ethanol, propanone (acetone), salt solution [a 0.1 wt% salt (NaCl) solution has been used as a solvent to separate food dyes by paper chromatography].

Safety notes
See page 11. No naked flames.

Curriculum links
Chromatography, acids and alkalis, indicators, dyes, polar/non-polar solvents.

Possible approaches
Paper chromatography of the coloured pens. Colour changes are due to pH changes of dyes/indicators (Vogel's Textbook of "Quantitative Inorganic Analysis" Fourth edition, p240-241 gives a good list of indicators). The "Magic" pen contains alkali. One approach might be to separate colours from a "funny" felt pen using paper chromatography, then dry the chromatogram (a hairdryer is handy) and dab the 'magic' pen on the separated colours to see which of the colours it affects.

☞ Perhaps put out acid and alkali as resource chemicals as a hint to lead students into thinking that the problem involves acid/base chemistry.

Extension work
Students could try to identify the names of indicators in each pen – question involving pH charts.

Are 'ordinary' felt pens susceptible to the magic pen?

THE ROYAL
SOCIETY OF
CHEMISTRY

Your notes

THE ROYAL
SOCIETY OF
CHEMISTRY

You are provided with one orange and one lemon.

☞ Your task

Find out which of the two contains the greater total amount of acid. You have laboratory glassware and equipment available, but no chemicals other than those which you are able to extract from the plant material provided. You are also provided with the ash from a charcoal barbecue or garden bonfire, which is rich in potassium carbonate.

▲ **HINT:** It is unwise to use all of your sample.

Based on a suggestion by P. Borrows.

THE ROYAL
SOCIETY OF
CHEMISTRY

THE ROYAL
SOCIETY OF
CHEMISTRY

Age	16 years upwards.
Time	100 minutes. Needs time before practical session to plan.
Group size	2–3.
Equipment & materials	Eye protection.

General

Beakers, burettes, pipettes (& safety fillers), graduated volumetric flasks, conical flasks, measuring cylinders, glass droppers, plastic syringes. Spatulas, glass stirring rods, watch glasses, test tubes, filter funnels and papers. Glass juicers, fine-mesh sieves, knives, white tiles to cut fruit on. Bunsen burners, tripods, gauzes, heatproof mats, clampstands. A centrifuge and balances should be available.

THE 'ASHING' PROCESS: (ash prepared before lesson by technician).

A large quantity of leaves, seaweed or other plant material can be 'ashed' to provide an alkali (largely sodium carbonate from sea plants, and potassium carbonate from land plants). [Some investigation may be needed to determine suitable plants & suitable quantities. See Annales de Chimie, xix, 157 and 194.]

Apparatus needed for 'Ashing'
Container for ashing, a large bulk of plant material (*eg* catering size coffee tin), Bunsen burner, tripod and heatproof mat. Access to a fume cupboard needed as ashing is smelly. 'Ashing process' likely to be very time-consuming!

NB Alternative ashes!
Ash from charcoal barbecue or garden bonfire/cigarette ash/'synthetic ash' = potassium carbonate.

Per group
Pestles and mortars – as big as possible (1 for each group).

Plant material
Unpickled beetroot or red cabbage leaves as indicator. 1 orange and 1 lemon of approximately equal size. 25 g of 'ash'(see ashing process or list of alternative ashes). Solvent.
Hot water.

Safety notes	See page 11. Ashing must be done in the fume-cupboard.
Curriculum links	Acids and alkalis, indicators, neutralisation.
Possible approaches	Note that graduated apparatus not needed to answer problem as stated (burettes and volumetric flasks are great distractors). Approaches already noted:-

1   The ash is dissolved in water to make an alkaline solution, this is then filtered. The oranges & lemons are sliced up and all the juice

THE ROYAL
SOCIETY OF
CHEMISTRY

squeezed out of them; the juice and chopped pieces are boiled in water and then filtered. The red cabbage leaves are boiled in water (the dye goes into solution), the red cabbage solution is filtered. A titration is then performed with both the acids from the lemon and the acids from the orange. The cabbage dye is used as an indicator (end point: red to green). The alkaline solution from the ash is placed in the burette.

Quantitative result
Typically lemons contain approximately 3.5 times more 'acid' than oranges. This result can be linked with taste.

2   Alternatively, an acid and a carbonate react to give off CO_2. This emission of CO_2 will result in a weight loss. By calculating this weight loss the relative amount of CO_2 given off can be calculated. As more acid will react to give off more CO_2, the quantity of CO_2 emitted can be used to determine which contains more acid.

Evaluation of solution

Credit could be given for:

Appropriate technique for preparing alkali solution;
preparing indicator;
extracting juice from fruit;
suitably diluting & titrating;
comments, explanation, presentation, accuracy.

Extension work

☛   Does acidity depend on variety of fruit, age, ripeness?
☛   Is peel different from flesh? Titrate separately.
☛   Experiment could also be allied to some work on wine-making (the wine producer needing to know the acidity of the grape).

Your notes

....

....

....

....

....

....

....

....

THE ROYAL
SOCIETY OF
CHEMISTRY

☞ Your task

Dye a piece of cloth in as many colours as possible using natural materials.

▲ The cloth will be tested for fastness (to cold and hot water), brightness of colour and variety of colours produced.

Note
▲ Wear an overall !

Based on a suggestion by J. Crellin.

THE ROYAL
SOCIETY OF
CHEMISTRY

THE ROYAL
SOCIETY OF
CHEMISTRY

Age	11 years upwards. (All abilities.)
Time	60 minutes to dye the cloth after planning. This experiment is best carried out during autumn.
Group size	2–4.
Equipment & materials	Eye protection.

Suggest students are warned in advance to bring an old shirt or a CDT apron to the session.

General
Beakers (all sizes), pestles and mortars, glass stirring rods, tongs, string. Buckets for rinsing cloth. Bunsen burners, tripods, gauzes, heatproof mats, clampstands.

Undyed woollen cloth (appeal for a very old blanket). Alternatively cotton may be used.

A selection of plant material, *eg* onion skins, red cabbage, beetroot, rhododendron leaves, acorns, used coffee grounds, pine cones, blackberries, redcurrants.

Mordanting salts: *eg* potassium aluminium sulphate(alum), iron(II)sulphate, copper(II)sulphate, tin(II)chloride.

Safety notes	See page 11. Warn students about danger from boiling liquids and steam.
Curriculum links	Dyeing. Mordants. Indicators.
Possible approaches	Dyeing can be used to supplement a unit on plants or one on colour & light, or it could be part of a cross-curricular unit on ancient civilisations. The dye is extracted by simmering a large quantity of the plant (flowers, berries, leaves or bark) in water. To create a stronger bond between the dye and the material, dyers often use mordants (*ie* fixing agents). As well as helping the dyes stick to the fibre, mordants also increase the colour range of the dye (many dyes give different colours with different mordants). Copper or iron pots may also act as mordants and affect the colour of the dye. You could add cream of tartar (acid potassium tartrate) to your materials list – it is used as an additive to brighten the colours.

NB There is a large quantity of stained glassware to wash up and some dyes are very difficult to remove.

Evaluation of solution	Credit could be given for:-

Fastness (to cold and hot water), brightness of colour and variety of colours produced. (Judges could also select their own criteria.)

Extension work	It is fun to dye other fabrics apart from wool, *eg* silk, cotton *etc*. Staple pieces of the various cloths together and immerse in the dye-bath.

Natural dyes work best with natural fibres. Find out about synthetic dyes (1850s onwards). Students use dyed silk in various art projects: *eg* make a colourful scarf or tie using the knowledge they have gained. A T-shirt promoting chemistry could be produced.

Colour fastness: Investigation of whether coloured cloth affected by light, heat, chlorine (important for swimwear) or perspiration?

References
KG Ponting "A Dictionary of Dyes and Dyeing" (London: Bell & Hyman Ltd., 1981); Jean Fraser "Traditional Scottish Dyes and how to make them" (Canongate Publishing Ltd., 1985); Lesley Bremness "The Complete Book of Herbs" – Chapter on Herbal dyes (London: Dorling Kindersley Ltd., 1988).

Your notes

....

....

....

....

....

....

....

....

....

....

....

....

....

....

....

....

THE ROYAL
SOCIETY OF
CHEMISTRY

☞ Your task

Design and build an electrical battery, which will give the greatest voltage. The battery is to be constructed using materials only likely to be found in a kitchen.

▲   A diagram of your invention, with some explanation, will be needed.

▲   Your final battery must be ready to be connected to the voltmeter used by the judges, at the appropriate time.

Based on a suggestion by P. Borrows.

THE ROYAL
SOCIETY OF
CHEMISTRY

THE ROYAL
SOCIETY OF
CHEMISTRY

Age	14 years upwards. (Given some guidance, younger students can do this as a 'fair test' – "Which combination of metals/electrolytes gives best voltage".)
Time	60 minutes (this may be rather short to allow for full investigation).
Group size	2–4.
Equipment & materials	Eye protection. Items from the 'junk' list, for example plastic egg container with 6 hollows for cells in series, (see page 20) – to encourage creativity. **General** A voltmeter (or multimeter) with 2 wires fitted, and crocodile clips at the other ends. (All the meters used should be of similar input resistance.) Additional leads and clips so that a number of cells can be constructed. Bulb. Something abrasive such as wire wool or glass paper to break oxide layers on the metal surfaces used as electrodes. The following additional items (**NB** this is not a final list, and not all of these items are required for a solution. This list can be varied:- Electrodes: carbon electrodes(pencils)/all sorts of metal cutlery (silver spoons *etc*)/off-cuts from copper pipes/brass curtain rings/aluminium foil or milk bottle tops/paper clips/pairs of scissors/pen tops. Materials to try as 'ionic solution' in cell: Bleach (do NOT use this with younger students)/bicarbonate of soda/sodium chloride/washing powder/washing soda(bath salts)/vinegar/lemonade/coca cola/couple of lemons/milk powder/ sugar/flour. Students could bring in a selection of solutions from home.
Safety notes	See page 11. Bleach could be hazardous if full strength. Safety warning of chlorine gas being given off when mixing vinegar and bleach.
Curriculum links	Simple electrolytic cells. Reactivity series. (16+ years could use salt bridges *etc* 13–14 year olds could use 2 dissimilar metals in a single electrolyte.)
Possible approaches	(The instructions are adequate for "high ability" groups but students of lower ability may need 'clues' to set them on the right course.) There could be three investigations:-

(i) which two metals are best? (Reinforcing the reactivity series.)

(ii) which electrolyte?

(iii) is it necessary to arrange for large area of electrodes or a "pile" effect (cells in series)?

For greatest voltage, just continue adding cells in series. With a restricted range of (safe) chemicals, some teachers think this could be used with junior school students – they would need to be shown how to use the voltmeter, but after that it would be a matter of trial & error, controlling variables *etc.*

Are students going to connect voltmeter directly to the cell?

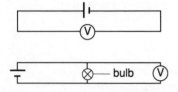

NB This is bad physics, as potential difference should be across a load, *eg* Many students don't realise that even if the current is insufficient to light up the bulb, there can still be a potential drop across it.
Another issue is polarisation of the electrodes. Do you take the initial surge voltage, or that after 10 seconds, or ...? These are problems that you should be aware of when judging the experiment.

Evaluation of solution

These are suggestions only:

1 No more than twenty cells may be connected in series.

2 At the time of the judging, the group should disconnect the wires from their voltmeter, and connect them instead to the judge's voltmeter.

3 The circuit will be complete, and the voltage reading after ten seconds recorded.

4 The winner is the group whose battery achieves the highest voltage. (Or alternatively, the winner is the group with the greatest voltage per individual cell – in which case the student sheet should be changed accordingly.)

5 In the event of a tie, the judge should take account of the non-messy "elegance" of the battery and its ability to keep a high potential under a moderate resistive loading.

Your notes

....

....

....

....

....

....

....

....

....

THE ROYAL
SOCIETY OF
CHEMISTRY

☞ **Your task**

Design and build an electrical battery, which will give the highest current. The battery is to be constructed using materials only likely to be found in a kitchen.

▲ A diagram of your invention, with some explanation, will be needed.

▲ Your final battery must be ready to be connected to the ammeter used by the judges, at the appropriate time.

Based on a suggestion by P. Borrows.

THE ROYAL
SOCIETY OF
CHEMISTRY

THE ROYAL
SOCIETY OF
CHEMISTRY

Age	14 years upwards. (Could be used for 13–14 year olds if they know about simple cells or were given a clue leading on from work on the reactivity series.)
Time	60 minutes (this may be rather short to allow for full investigation).
Group size	2–4.
Equipment & materials	Eye protection. Items from the 'junk' list (see page 20) – to encourage creativity. **General** A milliammeter (or multimeter) with 2 wires fitted, and crocodile clips at the other ends. Additional leads and clips so that a number of cells can be constructed. Bulb. Something abrasive such as wire wool or glass paper to break oxide layers on metal surfaces used as electrodes. The following additional items (**NB** this is not a final list, and not all of these items are required for a solution. The list can be varied as the judges think fit):- Electrodes: Carbon electrodes(pencils)/all sorts of metal cutlery/off-cuts from copper pipes/brass curtain rings/aluminium foil. Materials to try as 'ionic solution' in cell: Bicarbonate of soda/sodium chloride/washing powder/washing soda(bath salts) /vinegar/lemonade/coca cola/ bleach/couple of lemons/milk powder/sugar/flour. Students could bring in a selection of solutions from home.
Safety notes	See page 11. Bleach could be hazardous if full strength. Safety warning of chlorine gas being given off when mixing vinegar and bleach.
Curriculum links	Simple electrolytic cells. Reactivity series. (16+ years could use salt bridges *etc*. 13–14 year olds could use 2 dissimilar metals in a single electrolyte.)
Possible approaches	(The instructions are adequate for "high ability" groups but students of lower ability may need 'clues' to set them on the right course.) There could be three investigations:-

i   which two metals are best? (Reinforcing the reactivity series.)

ii   which electrolyte?

iii   is it necessary to arrange for large area of electrodes or a "pile" effect (cells in series)?

This is a very successful problem and always produces results. It can be taken at many different levels, from pure trial & error, to sophisticated chemical thinking. As set out, it asks for the highest current, which is rather more sophisticated than the greatest voltage, because it raises issues of the internal resistance which is probably suitable for 16+ years who could appreciate the theoretical issues. 14–16 year olds would get

THE ROYAL
SOCIETY OF
CHEMISTRY

a better start on the basis of voltage (see Kitchen potential experiment).
The best electrolyte is bleach, but as this is hazardous it could be
omitted with younger students. (With bleach you can get up to 0.5 A,
without bleach currents seem to be in the milliamp range.) Constructing
a battery of cells is a worthwhile technique. With a restricted range of
(safe) chemicals, some teachers think this could be used with junior
school students – they would need to be shown how to use the
milliammeter, but after that it would be a matter of trial & error,
controlling variables *etc*.

For cells connected in series: (Assume wires have no resistance.)

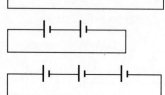

$$I \, (current) = \frac{V}{r} \quad \begin{array}{l} (voltage) \\ (r=internal \ resistance) \end{array}$$

$$I = \frac{2V}{2r}$$

$$I = \frac{3V}{3r}$$

Current stays the same (*ie* it is no use putting cells in series in an
attempt to increase the current).

For cells connected in parallel:
Voltage is same. (Assume wires have no resistance.)

Total resistance (R) of parallel 'network'

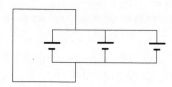

$$= \frac{r}{n} \quad \begin{array}{l} (internal \ resistance \ of \ cell) \\ (number \ of \ cells) \end{array}$$

The greater the number of cells, the smaller the resistance of the parallel
network, therefore the larger the current.

NB In practise, given the limitations of the 'junk' apparatus, it is difficult
to get a number of cells that are all identical.

Evaluation of solution   These are suggestions only:

1   At the time of the judging, the group should disconnect the wires
     from their ammeter, and connect them instead to the judges'
     ammeter.

2   The circuit will be complete, and the peak current is recorded.

3   The winner is the group whose battery achieves the highest "peak
     current".

4   In the event of a tie, the judges should take account of the
     non-messy "elegance" of the battery.

THE ROYAL
SOCIETY OF
CHEMISTRY

You are a group of survivors of a shipwreck on a small uninhabited island in the Pacific Ocean. Fortunately you have managed to salvage the ship's portable radio transmitter. However, you find that you are unable to send a SOS signal because the radio requires a small piece of lead to complete one of its circuits.

Plastic containers of acid from the ship's hold have been washed up on the beach. Also there appears to be an outcrop of lead ore on the island.

☞ **Your task**   Produce enough metallic lead that can be pressed into the gap so that the circuit can be completed and your SOS message can be sent.

▲   A diagram of your invention with, some explanation, will be needed.

Based on a problem used at the Norwich Chemical Olympiad 1984.

....

....

....

....

....

....

....

....

....

....

....

....

....

....

....

....

....

....

....

THE ROYAL
SOCIETY OF
CHEMISTRY

Age	15 years upwards. (Students need a prior knowledge of electrolysis.)
Time	60 minutes.
Group size	2–3.
Equipment & materials	Eye protection. Items from the 'junk' list (see page 20).

General
Pencil leads (for use as graphite electrodes), wire, crocodile clips, power pack (for radio transmitter – **NB** required to send message once gap is bridged).
Water, dilute nitric, hydrochloric, sulphuric acids.

Per group
100 g "lead ore" – a 50:50 mixture of lead nitrate and sand.

Safety notes	See page 11.
Curriculum links	Electrolysis. Raw materials.

Possible approaches
Students produce metallic lead by electrolysis from a given lead ore by improvising an electrolysis apparatus. Alternatively, students could separate the lead nitrate from the sand (*ie* by dissolving in water), evaporate, heat to lead oxide (caution NO_2 given off), then reduce with carbon.

Evaluation of solution
These are suggestions only:

1 Only the materials provided should be used to obtain the lead.

2 If the experiment is used as a competition the winner is the group that first produces enough lead to be pressed into a gap in an electrical circuit, in order to complete the circuit (light bulb, ring a bell *etc*).

Extension work
Students investigate ways of producing more lead during electrolysis, *eg* increase the area of the electrodes in the solution, put the electrodes closer together, increase the concentration of the solution, leave it for longer, increase the voltage and therefore the current.

Your notes

....

....

....

....

....

THE ROYAL
SOCIETY OF
CHEMISTRY

Esso

THE ROYAL
SOCIETY OF
CHEMISTRY

A small local company is considering re-cycling the
chemicals from dead zinc-carbon batteries and has
approached your school for advice on how many different
chemicals are contained in zinc-carbon batteries.

The company has provided your school with a number of
dead zinc-carbon batteries which they have cut in half. They
have also provided new batteries and salty water.

☞ **Your task**

Find out how many chemicals you can obtain from the dead
batteries.

Based on a suggestion by P. Borrows.

THE ROYAL
SOCIETY OF
CHEMISTRY

THE ROYAL
SOCIETY OF
CHEMISTRY

Age	15 years upwards. (Students need a prior knowledge of electrolysis but any reference books may be used.)
Time	A pre-set problem. Approximately 90 minutes.
Group size	2–4.
Equipment & materials	Eye protection.

General
Electrolysis cells. Wires, crocodile clips, carbon electrodes. A selection of standard laboratory apparatus. Newspaper to place on bench when dissecting cell (messy!). On the day display boards and tables may be required for presentations. Paper.
Sodium chloride, distilled water.

Per group
Cells for electrolysis, one dead cell PRECUT in half (most students could provide a run-down cell for themselves).

NB The cells must be of the zinc-carbon type, *eg* Ever Ready Blue Seal, Boots SP.

NB Other chemicals should be made available for testing/confirming purposes only, *eg* aqueous hydrogen peroxide (20 volume) – about 10 cm^3 per group (as test for MnO_2). Dilute nitric, hydrochloric acids. Dilute sodium hydroxide. Dilute silver nitrate solution. Indicator papers, splints.

Safety notes	See page 11. ONLY zinc-carbon batteries should be used (alkaline type batteries should NOT be used).

Zinc-carbon batteries contain traces of mercury. Students must wash their hands after the practical.
Caution against students wanting to try to initiate a hydrogen-chlorine explosion to synthesise HCl(g) – perhaps issue a safety warning concerning this.
Warning of sharp edges on battery casing.

Curriculum links	Electrolysis. Chemical reactions.
Possible approaches	Use this experiment as a long-term class project. Students may use books to plan out their work and confirm their findings in the laboratory. Groups of students design their own investigations.

Initially groups should carry out the electrolysis of salt water using the cells provided. At the anode; if the NaCl solution is concentrated, then chlorine is given off (test: chlorine bleaches moist litmus paper); if the NaCl solution is dilute, oxygen is given off (if O_2 reacts with the carbon electrode then CO_2, $NaHCO_3$, and Na_2CO_3 will be formed). At the cathode: hydrogen is given off. The solution left in the beaker is alkaline (NaOH).

THE ROYAL
SOCIETY OF
CHEMISTRY

Electrolysis is very slow with only 3 volts available, students will find it quicker to use more batteries for electrolysis.

After carrying out electrolysis, the dead pre-cut batteries may then be used as new materials for further processing. The carbon rod and zinc case are easy to identify. Take out and clean the carbon rod. The black insoluble material around the carbon rod is a mixture of manganese dioxide and powdered carbon. The outer cell filling is a mixture of ammonium chloride and zinc chloride.

$$NH_4Cl + NaOH \longrightarrow NH_3 + NaCl + H_2O$$
$$\text{Heat } NH_4Cl \longrightarrow \text{may get some HCl separating.}$$
$$ZnCl_2 + NaOH \longrightarrow Zn(OH)_2, Na_2ZnO_2$$

References

Science and Technology in Society units: SATIS 7, No. 706 'Dry Cells' (ASE, 1986).

Evaluation of solution

Credit should be given for each separate element/compound prepared or isolated. It is important to have a clear idea as to the criteria to judge whether or not a compound has been prepared. Credit should be given for the presentation of the display and for the ingenuity in working out the chemistry.

The results could be presented in the form of a flow-chart on a poster. Where possible, students should present evidence that they have actually made the chemicals mentioned.

Your notes

THE ROYAL
SOCIETY OF
CHEMISTRY

☞ Your task

Your school has been challenged by the promoters of "Energy-Saving" week in your local area to cook an egg over a nightlight, using a tripod and aluminium foil. They want to see which group uses the least energy to cook their egg.

Based on a suggestion by S. Lindley.

THE ROYAL
SOCIETY OF
CHEMISTRY

THE ROYAL
SOCIETY OF
CHEMISTRY

Age	11–13 years. (All abilities.)
Time	30-60 minutes depending on the group.
Group size	2–3.
Equipment & materials	Eye protection.
	Per group A tripod, kitchen foil (limited to about 0.25 square metre at most), a nightlight or candle, heatproof mats, matches, spoons, an empty tin. A top pan balance. An egg, sodium chloride (flavouring!).
Safety notes	See page 11. Safety aspects of eating in the lab.
Curriculum links	Energy, insulation.
Possible approaches	Students are given little information about the factors which will affect the speed of cooking. They usually realise surrounding the flame with aluminium foil conserves energy. **NB** By doing so it is fairly easy to prevent oxygen getting to the flame so that it goes out! Covering the egg with foil decreases cooking time significantly. Eggs cook much faster if they are scrambled (mixed) first. If this experiment could be done in the Home 'Eggonomics' department students could actually eat their results (experiment has cross-curricular possibilities) – the results are often surprisingly edible. If the foil splits the reaction gets messy and smelly.
Evaluation of solution	The mass of the nighlight is measured before and after the experiment to see which group has used least energy.
Suggested write-up	Students could produce a poster for display at the 'Energy-Saving' week, showing the steps they took to produce their cooked egg. They might also include some of the actual materials they used.

Your notes

....

....

....

....

....

....

....

....

THE ROYAL
SOCIETY OF
CHEMISTRY

....

....

....

....

....

....

....

....

....

....

....

....

....

....

....

....

....

....

....

....

....

....

....

....

THE ROYAL
SOCIETY OF
CHEMISTRY

A timetable clash means your next chemistry lesson has to be in a classroom. Your teacher says you won't be able to do practical work as you won't be able to use Bunsen burners. Not wanting a double period of theory, you wonder whether you could use a candle instead. Would it produce a worthwhile amount of energy?

☞ **Your task**

Find the number of candles which have the heating power of one Bunsen burner.

Based on a suggestion by R.F. Kempa.

THE ROYAL
SOCIETY OF
CHEMISTRY

THE ROYAL
SOCIETY OF
CHEMISTRY

Age	11–13 years. (Higher ability 11–12 year olds, the less able will need guidance. Could also be used with lower ability 14–15 year olds.)
Time	50–70 minutes depending on ability.
Group size	2–3.
Equipment & materials	Eye protection.
	General Accurate balances should be available.
	Per group Candle, thermometer, small metal cans, glass beakers (100 cm³), measuring cylinder (10, 50 cm³), stopclock. Bunsen burner, tripod, gauze, heatproof mat, clampstand.
Safety notes	See page 11. **NB** Large pieces of wick cause very large flames (this requires careful supervision of the 'bright sparks'!).
Curriculum links	Energy.
Possible approaches	Students should compare the effect of heating the same amounts of water, using a candle and a Bunsen burner, over the same time period, *eg* 2, 5 or 10 minutes. Many groups fail to appreciate that it is the temperature change that is important, *ie* they try to find an answer simply by comparing final temperatures. Other factors may be considered, such as the suitability of different heat sources, *eg* sooty candle flame.
Extension work	The experiment indicates to students the number of candles equivalent to a Bunsen burner. A simple extension of this would be to devise an experiment to test that hypothesis. (The results from one school suggest that 2 candles = 1 Bunsen burner. If this is true why don't all schools use candles?)

Your notes

....

....

....

....

....

....

....

THE ROYAL
SOCIETY OF
CHEMISTRY

THE ROYAL
SOCIETY OF
CHEMISTRY

It is sometime in the 21st century, fossil fuels have all but run out and all forms of heating are unimaginably expensive.

Some students lost on a Department of Education and Science expedition find an old waterproofed lime kiln with calcium oxide (quicklime) intact. Conscious that salmonella poisoning is still very much a possibility they are anxious to cook some eggs they have found ...

☞ Your task

You are provided with one egg which you are required to cook using a chemical reaction.

Before cooking the egg, have your proposed method checked.

WARNING
▲ Calcium oxide can cause burns, wear eye protection.

Based on a suggestion by V. Herbert/R.D. Langstaff.

THE ROYAL
SOCIETY OF
CHEMISTRY

THE ROYAL
SOCIETY OF
CHEMISTRY

Age	12–14 years.
Time	Approximately 50 minutes, excluding introduction and evaluation.
Group size	2–3.
Equipment & materials	Eye protection.

Items from the 'junk' list, including insulation (see page 20) – to encourage creativity.

General
Aluminium foil, plastic spoons, glass stirring rods, glass beakers, boiling tubes, test tubes, thermometers.

Calcium Oxide: FRESH commercial CaO is best (most bottles of CaO are actually $Ca(OH)_2$ and therefore useless). Alternatively CaO can be obtained by roasting $CaCO_3$ in kiln (requires a temperature over 1000 °C), some schools may have a Muffle furnace (**NB** a Bunsen burner is not hot enough). The CaO should be tested before the session, otherwise disappointment may result.

Per group
An egg (as small as possible), fresh calcium oxide (approximately 50 g per egg), water.

Safety notes

See page 11. A STRONG SAFETY WARNING – 'Calcium oxide can cause burns', students must wear eye protection.

☞   One student added 5 cm^3 of water to about 50 g of CaO in one go and the evolution of heat was very rapid – cracking the pyrex beaker quite violently. Also the reaction seems to have an induction period (probably $Ca(OH)_2$ on surface), one can add quite a bit of water with seemingly no effect, then the container becomes too hot to touch quite quickly.

Because of salmonella risk insist on strict hygiene, wash hands at end *etc.*

NO EATING THE EGGS!

Curriculum links	Exothermic reactions.
Possible approaches	Each group must cook an egg using only the energy generated by the reaction of calcium oxide + water.

The same amount of calcium oxide could be provided to each group. The degree of cooking will then be a reflection of the design qualities.

Some groups may need guidance on how to produce the energy and how to get the maximum heat from the reaction.

Beware – the energy generated by the addition of even small amounts of water to CaO may melt plastic containers.

All students can succeed to a certain extent – whether egg is raw or hardboiled. It may be necessary to demonstrate that you do produce heat by mixing CaO and water. You might also suggest that all groups investigate the basic CaO/H_2O process on a small scale before they attempt to cook their eggs.

Have a grand 'CRACKING OPEN' session at end of lesson

Evaluation of solution   These are suggestions only:

1   Before cooking the egg groups should describe their proposed method to the judges.

2   No form of heating, other than chemical, is to be used for the cooking process.

3   A cooked egg is one in which the white is firm. The yolk may or may not be runny. The judges's decision is final.

4   The winner is the group which produces a cooked egg within the given time.

5   If there are no fully cooked eggs, then the group with the most cooked egg wins.

6   If there are a number of cooked eggs, then the design of the apparatus and the amount of calcium oxide used could be taken into account.

Extension work   Design a 'hot can' that could be used for cooking foods on a future expedition. Discuss the possible uses of portable heat sources.

Your notes

....

....

....

....

....

....

....

THE ROYAL
SOCIETY OF
CHEMISTRY

You are going on a weekend camping expedition. During the expedition you will have to do all your own cooking and therefore you have to carry the cooking fuel with you. You have a choice of two fuels, 'Lotahot' or 'Superheat'.

☛ **Your task**

Decide which is the better fuel to use for boiling water and hence which fuel to take with you on the expedition. You will need to consider two things:

1   How long it takes to boil water.

2   The quantity of fuel needed.

Have your method checked for safety, before you try your experiment.

Based on a suggestion by R.F. Kempa.

THE ROYAL
SOCIETY OF
CHEMISTRY

THE ROYAL
SOCIETY OF
CHEMISTRY

Age	12–16 years. (All abilities. Could be used for younger students as a 'fair test' exercise.)
Time	Discussion and planning approximately 30 minutes. Practical – 60 minutes.
Group size	2–3.
Equipment & materials	Eye protection.

General
Old tin cans, beakers, thermometers, tripods, gauzes, heatproof mats, measuring cylinders, clampstands, woodblocks, aluminium foil, metre rulers, stopclocks, top-pan balances.

Per group
Fuels – two solid fuels, *eg* 'meta' tablets and paraffin/candle wax.

Safety notes	See page 11. Ensure students use only small quantities of fuel, *ie* NO large containers of fuel. **NB** Warn students that the fuel containers will become very hot – therefore do not touch.
Curriculum links	Energy.
Possible approaches	Old small cans are very good for this experiment as they are easy to come by, and can be regarded as disposable.
Suggested write-up	Students write a report for the "Consumer watch-dog chemistry column" in a consumer magazine.
Evaluation of solution	These are suggestions only:

1   Scientific method, *ie* isolation of possible variables, changing one variable at a time (*eg* distance of flame from beaker).

2   Satisfactory solution achieved, *eg* burning all the fuel is not a good method.

3   Reasoning about which is best fuel to take on the expedition.

Extension work	How would you improve this experiment if you were doing it again?

Your notes

....

....

....

....

....

**THE ROYAL
SOCIETY OF
CHEMISTRY**

THE ROYAL
SOCIETY OF
CHEMISTRY

☞ **Your task**

You are on a camping expedition and need to make a simple meal consisting of one cup of coffee, one boiled egg and a portion of rice pudding.

Based on a suggestion by R. Lewin.

THE ROYAL
SOCIETY OF
CHEMISTRY

THE ROYAL
SOCIETY OF
CHEMISTRY

Age 12–16 years. (All abilities.)

Time 180 minutes.

Group size 2–3.

Equipment & materials Eye protection.

General
Aluminium foil.

Per group
1 metal coat hanger, copy of the Times Education Supplement or similar, 1 nightlight, 1 standard candle, 1 box of matches. Cup, plate, spoon, small tools including pliers, scissors.

Egg, milk, rice, butter, sugar, coffee.

Formula for 1 portion of rice pudding: 3/4 tablespoon short grain rice, 1/2 tablespoon of caster sugar, 1/4 pint milk (150 cm^3), shavings of butter.

Wash rice and put into a buttered dish with the sugar. Pour on the milk, top with shavings of butter.

Safety notes See page 11. Remember fire is a hazard, and the safety aspects of eating in the laboratory.

Curriculum links Energy, insulation.

Possible approaches Candles can be cut into pieces. Covering the egg with foil decreases cooking time significantly.

Evaluation of results These are suggestions only:

1 All products to be tested when ready.

2 Credit should be given for quality of products.

3 Elegance of design.

Your notes

....

....

....

....

....

THE ROYAL
SOCIETY OF
CHEMISTRY

THE ROYAL
SOCIETY OF
CHEMISTRY

☛ ## Your task

Design and make a device (with final dimensions NOT exceeding 15 cm x 15 cm x 15 cm) which will sink in water, and then after a reasonable length of time, rise to the surface. As far as possible, the device is to be constructed from 'junk' materials.

▲ The final device must be loaded with chemicals, and be ready to start the experiment when told. You cannot do anything to the device, *eg* add anything to it (or to the water) once it is in the tank.

▲ There must be no connection between the device and the surface, once it has sunk.

Based on a suggestion by Sussex SATRO.

THE ROYAL
SOCIETY OF
CHEMISTRY

Esso

THE ROYAL
SOCIETY OF
CHEMISTRY

Age	10 years upwards.
Time	It is suggested that either:-

an entire morning be devoted to the problem (*eg* on the last day of term), which would allow 2 hours for practical activities and 30 minutes for judging

or

the problem be given to the class as a homework exercise 2 weeks or so before the judging. Judging could then take place in a normal double science lesson, allowing 45 minutes for repair and final adjustments, and 30 minutes for judging.

(The exercise is better as a pre-set problem for younger students.)

Group size	3–4.
Equipment & materials	Eye protection.

Items from the 'junk' list, for example thin-necked plastic lemonade bottles, yoghurt pots, plastic 'specimen' containers with lids, plastic bags, balloons, scissors (see page 20) – to encourage creativity.

Test tank, containing water to a depth of 20 cm.

Judges will require a stopclock and a 30 cm ruler plus marker pen.

General
Corks to fit yoghurt pots, plastic beakers (100 cm^3), stopclocks, 10 g masses, rubber bands, rubber tubing (1 metre), sticky tape or masking tape, string, plastic buckets for 'test runs'.

Selection of chemicals: Salt, sugar, tartaric acid, marble chips, hydrochloric acid (2 mol dm^{-3}), 'Alka Seltzer' tablets, Andrews' liver salts, or a mixture of solid sodium hydrogencarbonate/citric acid (1 teaspoon of sodium hydrogencarbonate to 3 teaspoons of citric acid).

Access to water.

Per group
Identical teaspoons (can be plastic). A 30 cm ruler.

Safety notes	See page 11.
Curriculum links	Solubility. Reaction of weak 'solid' acids with sodium hydrogencarbonate in the presence of water.
Possible approaches	Generating a gas which causes a container to become buoyant; weighting a container with soluble material and allowing water to dissolve it so that the container then rises to the surface. (The sinking or floating/rising aspects could be investigated separately.)

THE ROYAL
SOCIETY OF
CHEMISTRY

Evaluation of solution   These are suggestions only:

1   Students should not do anything to the device, *eg* add something to it (or to the water) once it is in the tank. Also, there must be no connection between the device and the surface, once it has sunk.

2   The final device must be loaded with chemicals, and be ready to start the experiment when the judge says so.

3   When placed on water the device should float for at least 5 seconds. Then when left it should sink until it is at least 5 cm under the surface of the water. It must stay at least 5 cm under water for any reasonable length of time, but must then rise to the surface where it should remain afloat again for another 5 seconds.

4   Judges should be looking for simplicity, reliability, elegance and humour.

Extension work   Increasing the complexity of the behaviour required in terms of time limits and depth limits. Extend to actual lifting of a submerged object.

Your notes

....

....

....

....

....

....

....

....

....

....

....

....

....

....

THE ROYAL
SOCIETY OF
CHEMISTRY

☞ Your task

Design and make a device to lift an Oxo cube as high as possible using, as an energy source, the reaction between 1 level teaspoon of bicarbonate of soda (sodium hydrogencarbonate) and 3 level teaspoons of citric acid. Once lifted, the Oxo cube must stay there. As far as possible, the lifting device is to be constructed from 'junk' materials.

▲   Your final device must be loaded with chemicals, and be ready to start the experiment when the judges say so.

Based on a suggestion by P. Borrows.

THE ROYAL
SOCIETY OF
CHEMISTRY

THE ROYAL
SOCIETY OF
CHEMISTRY

Age	12–18 years.
Time	It is suggested that either:-

an entire morning be devoted to the problem (*eg* on the last day of term), which would allow 2 hours for practical activities and 30 minutes for judging

or

the problem be given to the class as a homework exercise 2 weeks or so before the judging. Judging could then take place in a normal double science lesson, allowing 45 minutes for repair and final adjustments, and 30 minutes for judging.

(The exercise is better as a pre-set problem for younger students.)

Group size	3–4.
Equipment & materials	Eye protection.

Items from the 'junk' list, for example plastic syringes, (see page 20) – to encourage creativity.

The judges will require a tape measure/metre rule.

Per group
Identical teaspoons (can be plastic).
Sodium hydrogencarbonate (maximum amount = 3 level teaspoons), citric acid (maximum amount = 9 level teaspoons), access to water.
Butter/margarine to reduce friction.

Safety notes	See page 11.
Curriculum links	Production of carbon dioxide gas.
Possible approaches	Perhaps guidance needed for younger age groups to say that water is needed for the reaction, or use 'Andrews'. The reaction might be used to do the lifting, or it could be used to start the lifting, *eg* to trigger movement of a counterbalance. One group solved the problem by using the displacement of water.
Evaluation of solution	These are suggestions only:

1   The final device must be loaded with chemicals, and be ready to start the experiment when the judge says so.

2   The judge will provide each group with the levelled teaspoons of chemicals for the test. (Judges may prefer to weigh out the relevant amounts.)

3   The winner is that device which lifts the Oxo cube the highest vertical distance, and keeps it there.

THE ROYAL
SOCIETY OF
CHEMISTRY

4 In the event of a tie, the judge should take into account the elegance of the solution, given the requirement that the device shall be constructed mainly from 'junk' materials.

'The Lift Oxo Cubes to Dizzy Heights Challenge Cup'! So far the highest lift is 22 cm. If your students beat this, write to the Royal Society of Chemistry (Education Department) with details. The highest lift will be published from time to time in 'Education in Chemistry'.

Extension work

To increase the chemical content the task could be extended by prior (or subsequent) experimentation, to select best choice of gases/chemicals.

Your notes

....

....

....

....

....

....

....

....

....

....

....

....

....

....

....

....

....

THE ROYAL
SOCIETY OF
CHEMISTRY

☞ Your task

Design, and make a boat. The boat is to be propelled by the reaction between 1 teaspoon of bicarbonate of soda (sodium hydrogencarbonate) and 3 teaspoons of citric acid. As far as possible, the boat is to be constructed from 'junk' materials.

The winner is that boat which travels the furthest distance.

▲ Your final device must be loaded with chemicals, and be ready to start the experiment when the judges say so.

Based on a suggestion by P. Borrows.

THE ROYAL
SOCIETY OF
CHEMISTRY

THE ROYAL
SOCIETY OF
CHEMISTRY

Age	13 years upwards.
Time	It is suggested that either:-

an entire morning be devoted to the problem (*eg* on the last day of term), which would allow 2 hours for practical activities and 30 minutes for judging

or

the problem be given to the class as a homework exercise 2 weeks or so before the judging. Judging could then take place in a normal double science lesson, allowing 45 minutes for repair and final adjustments, and 30 minutes for judging.

(The exercise is better as a pre-set problem for younger students.)

Group size	3–4.
Equipment & materials	Eye protection. Items from the 'junk' list (see page 20) – to encourage creativity. A testing tank: depends on what is available in your laboratory. However, the type of tank will determine how you evaluate the distance travelled by the boats (see possible approaches below).

Per group
Identical teaspoons (can be plastic).
Sodium hydrogencarbonate (maximum amount = 3 level teaspoons), citric acid (maximum amount = 9 level teaspoons), access to water.

Safety notes	See page 11.
Curriculum links	Production of carbon dioxide gas.
Possible approaches	A suitable long tank may be constructed from stout cardboard, lined with heavy duty polythene, or by using plastic guttering. The distance each boat travels can then be measured. Alternatively, a plastic washing-up bowl could be used. A clampstand is placed in the centre of the bowl and the boat attached to the clampstand by a piece of cotton, so that it is free to sail round the bowl. The number of times the boat goes round the clampstand is then measured. In summer months, you could use a children's paddling pool as an outdoor testing tank. (Some students will do some re-designing when they realise that in a washing up bowl the biggest boat is not the best!) "A good exercise with wide access across the age range". Younger age groups may need guidance about why water is needed for the reaction. The fuel could be carried on the boat or gas could be generated separately and stored in a balloon.
Evaluation of solution	These are suggestions only:

1   The final device must be loaded with chemicals, and be ready to start the experiment when the judges say so.

THE ROYAL
SOCIETY OF
CHEMISTRY

2   The judges will provide each group with the levelled teaspoons of chemicals for the test. (Judges may prefer to weigh out the relevant amounts.)

3   The boat will start with the rearmost structure at the end of the tank and the distance travelled will be measured from the end of the tank to the rearmost part of the boat or alternatively, the number of revolutions will be measured.

4   The winner is the boat which travels the furthest.

5   In the event of a tie, the judges should take into account the elegance of the solution, given the requirement that the device shall be constructed mainly from 'junk' materials.

Extension work

To increase the chemical content the task could be extended by prior (or subsequent) experimentation, to select best choice of gases/chemicals.

Your notes

....

....

....

....

....

....

....

....

....

....

....

....

....

....

THE ROYAL
SOCIETY OF
CHEMISTRY

☞ Your task

Design and make a device to lift as heavy an object as possible at least 10 cm using, as an energy source, the reaction between 1 level teaspoon of bicarbonate of soda (sodium hydrogencarbonate) and 3 level teaspoons of citric acid. Once lifted, the object must stay there. As far as possible, the lifting device is to be constructed from 'junk' materials.

▲   Your final device must be loaded with chemicals, and be ready to start the experiment when the judges say so.

Based on a suggestion by P. Borrows.

THE ROYAL
SOCIETY OF
CHEMISTRY

THE ROYAL
SOCIETY OF
CHEMISTRY

Age	14-18 years.
Time	It is suggested that either:- an entire morning be devoted to the problem (*eg* on the last day of term), which would allow 2 hours for practical activities and 30 minutes for judging or the problem be given to the class as a homework exercise 2 weeks or so before the judging. Judging could then take place in a normal double science lesson, allowing 45 minutes for repair and final adjustments, and 30 minutes for judging. (The exercise is better as a pre-set problem for younger students.)
Group size	3–4.
Equipment & materials	Eye protection. Items from the 'junk' list (see page 20) – to encourage creativity. The judges will require access to a top-pan balance. **Per group** Standard masses (10 g & 100 g slotted masses, and 1 kg masses for the ambitious!), a metre or half-metre ruler, identical teaspoons (can be plastic). Sodium hydrogencarbonate (maximum amount = 3 level teaspoons), citric acid (maximum amount = 9 level teaspoons), access to water.
Safety notes	See page 11.
Curriculum links	Production of carbon dioxide gas.
Possible approaches	Perhaps guidance needed for younger age groups to say that water is needed for the reaction. The reaction might be used to do the lifting, or it could be used to start the lifting, *eg* to trigger movement of a counterbalance.
Evaluation of solution	These are suggestions only:

1   Any device which is, or which appears to be, unsafe should be disqualified immediately.

2   The final device must be loaded with chemicals, and be ready to start the experiment when the judges say so.

3   The judges will provide each group with the levelled teaspoons of chemicals for the test. (Judges may prefer to weigh out the relevant amounts.)

4   Although "standard" masses will be provided, groups do not have to use these as their load. However, it must be possible to remove the load from the lifting device, in order for the judges to check its

mass. This must not destroy the lifting device. Thus, the mass of the device itself must not be counted towards the mass of the load.

5   The winner is that device which lifts the largest mass through at least 10 cm.

6   There is no penalty for lifting the object by more than 10 cm, as long as it stays at least 10 cm above its starting point.

7   In the event of a tie, the judges should take into account the elegance of the solution, given the requirement that the device shall be constructed mainly from 'junk' materials.

'The Heavy Lift Cup Challenge'! So far the best lift = 2 kilograms. If your students beat this, write to the Royal Society of Chemistry (Education Department) with details. The heaviest lift will be published from time to time in 'Education in Chemistry'.

Extension work

To increase the chemical content the task could be extended by prior (or subsequent) experimentation, to select best choice of gases/chemicals.

Your notes

....

....

....

....

....

....

....

....

....

....

....

....

....

THE ROYAL
SOCIETY OF
CHEMISTRY

Your task

Alka Seltzer tablets fizz when placed in water. Use this
reaction to make a device which will measure a time interval
of up to 30 seconds as accurately as possible.

Based on a suggestion by G. Woodford.

THE ROYAL
SOCIETY OF
CHEMISTRY

THE ROYAL
SOCIETY OF
CHEMISTRY

Age	9–13 years.
Time	120 minutes.
Group size	2–3.
Equipment & materials	Eye protection. Items from the 'junk' list (see page 20) – to encourage creativity.

General
See sketches of 'popular answers'.

Per group
A stop clock or equivalent for calibrating their device. These should be removed when judging.

2 Alka Seltzer tablets for investigating – groups will need a further tablet for the final measurement. (A mixture of solid sodium hydrogencarbonate/citric acid could be used as a substitute: one teaspoon of sodium hydrogencarbonate to 3 teaspoons of citric acid.)

Access to water.

Safety notes
See page 11.

Curriculum links
Handling and collecting gases. Idea of using water to indicate the presence/movement of an invisible gas.

Possible approaches
Students plan the experiment and if approved (even though it might not work) they should be allowed to continue and make a clock. One class who tried this experiment had no experience of handling and collecting gases. They needed considerable pointing in the right direction to enable them to get started.

Problem
Because most of the CO_2 gas initially produced dissolves in the water a major problem could be obtaining a linear scale. As a result some students might ask for a CO_2 source to pre-saturate the water.
"Having a mixture of sodium hydrogencarbonate and citric acid for preliminary trials saved a great deal on the cost of the Alka Seltzer tablets".

Popular answers

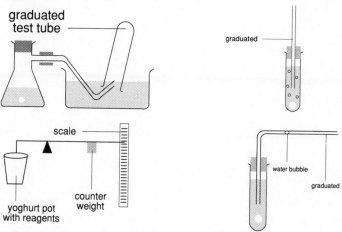

Evaluation of solution

These are suggestions only:

1   When it is time to judge the entries, all clocks will be removed. The judges will make a signal (*eg* with a lamp or bell), and repeat it not more than 30 seconds later. Groups must measure the time interval between the two signals.

2   The group with the most accurate measurement of the time interval is the winner.

Class winners could go forward into an inter-class final. This experiment is particularly suitable for a competition.

Extension work

Another possibility is using an indicator, *ie* 'alkali + indicator'. When enough CO_2 is produced to neutralise the alkali, then colour changes.

Your notes

....

....

....

....

....

....

....

....

....

....

....

....

....

....

THE ROYAL
SOCIETY OF
CHEMISTRY

☞ **Your task**

Determine which soft drink contains the most fizz.

Based on a suggestion by R.F. Kempa.

THE ROYAL
SOCIETY OF
CHEMISTRY

Age	13–16 years. (Depends upon their experience rather than age.)
Time	180–270 minutes.
Group size	2–4.
Equipment & materials	Eye protection. Items from the 'junk' list, for example, plastic bags, rubber bands, twist ties/string (see page 20).

General

Measuring cylinders, gas syringes, test tubes, bungs, conical flasks & bungs to fit them with a glass through tube (or side-arm flasks or side-arm boiling tubes), rubber tubing, washing up bowls, spatulas. Bunsen burners, tripods, gauzes, heatproof mats, clampstands. Balance to weigh full can.

Selection of solids: salt, sugar, sand, anti-bumping granules. Access to water.

Per group

Ice-cold cans of soft 'carbonated' drinks: two cans of each type of drink.

Safety notes	See page 11.
Curriculum links	Gases dissolving in water, solubilities of gases with temperature, adding a solid releases the gas.
Possible approaches	The wording does not require groups to produce maximum volume of gas, or even to measure volume at all. Students need to know that gases are less soluble in hot than in cold – students may 'know' this without being aware, *eg* why do drinks go 'flat' on a hot day (precursor – look at solubilities of gases with temperature), or that adding a solid releases the gas. Students may require help with ideas – collecting the gas, 'control', fair comparison *etc*. Greater guidance relating 'fizz' to volume of gas or mass would mean that 11+ students should be able to attempt it. Also it could be made more difficult by asking students to get all the gas out (see The "Real Thing" experiment).

Approaches already noted: CO_2 given off collected by displacement of water. (This is not as easy as it sounds – the gas produced dissolves in water until water is saturated.) Also loss of mass due to CO_2 given off.

☛ One teacher used a "Soda Stream" maker, so students (14–15 year olds) spent most of their time squeezing in different amounts of CO_2 into bottles of water, and then testing out their ideas on measuring the amount of dissolved gas. In the last lesson the teacher put different amounts of CO_2 in 2 bottles and students had to work out which contained most fizz using their method of recovery.

☛ "We did not attempt this activity as a race due to the fact that the range of abilities is great and I felt that it was important that everyone felt that they had achieved something. With a group of

THE ROYAL
SOCIETY OF
CHEMISTRY

similar abilities, with similar background knowledge *etc*, I would have been quite happy to use this material in the form of a competition".

☞ The group should demonstrate their method at end, not just give results.

Extension work   What is the gas? (Students must find a way of proving the gas is CO_2.)

"What happens if we increase the temperature"? (This could relate to river temperature, oxygen concentration and fish numbers.)

Your notes

....
....
....
....
....
....
....
....
....
....
....
....
....
....
....
....
....
....

THE ROYAL
SOCIETY OF
CHEMISTRY

☞ **Your task**

1   How much carbon dioxide will a stick of blackboard chalk produce?

or

2   Who can produce the largest amount of carbon dioxide from a stick of blackboard chalk?

Based on a suggestion by E. Grimble.

THE ROYAL
SOCIETY OF
CHEMISTRY

THE ROYAL
SOCIETY OF
CHEMISTRY

Age	14–18 years.
Time	60 minutes (perhaps up to 80 minutes for 14–16 year olds, depending on preparation).
Group size	2–3.
Equipment & materials	Eye protection. Students could put equipment lists in for the technician.

General

Conical flasks & bungs to fit them with a glass through tube (or side-arm flasks or side-arm boiling tubes), rubber tubing, thistle funnels, measuring cylinders, beakers, burettes, gas jars, troughs, beehive shelves, test tubes, boiling tubes, gas syringes, clampstands, mortars and pestles. Top pan balances should be available.
Access to water.

Per group

Students are given a bottle of dilute hydrochloric acid and one stick of white blackboard chalk (* note for judges:- check in the laboratory before practical that your chalk is the sort that gives appreciable amounts of CO_2). Students should be told that the amount of CO_2 given off is more than 100 cm^3 but less than 500 cm^3. They should be warned that CO_2 is soluble in water.

Safety notes	See page 11. Students must be warned about the handling of the acid. Powdered chalk reacts fast and corks pop if delivery tubes are constricted.
Curriculum links	Rates of reaction. Handling and collecting gases.
Possible approaches	Depending on ability and experience students are asked one of the two questions ((**i**). How much CO_2 will a stick of blackboard chalk produce? Or (**ii**). Who can produce the largest amount of CO_2 from a stick of blackboard chalk?) Experiment is useful in consolidating the students' knowledge of types of apparatus available.

Many students are not prepared for the large volume of CO_2 given off despite having a rough volume quoted! Approaches already noted: Collection in a gas syringe. Collection under water into beakers, burettes and measuring cylinders. Also mass loss using a balance.

It is difficult for the teacher to know the precise calcium carbonate content of the blackboard chalk since the boxes do not give a list of contents (blackboard chalk contains largely calcium sulphate, with calcium carbonate), but a fruitful discussion of the groups results should elicit all manner of fresh ideas as to why one method gave more than another. Sometimes the best results come from those groups that have produced the gas as quickly as possible using the whole stick, and collecting the gas in a gas syringe.

THE ROYAL
SOCIETY OF
CHEMISTRY

Extension work   Use coloured chalks to see if there are any differences. For 16–18 year olds:- Calculate the % $CaCO_3$ in the blackboard chalk from volume CO_2 given off and mass of stick of blackboard chalk.

Your notes

....
....
....
....
....
....
....
....
....
....
....
....
....
....
....
....
....
....
....
....
....
....

THE ROYAL
SOCIETY OF
CHEMISTRY

The Chemical Cola Company has just got hold of the secret
recipe for the "Real Thing". Before production can begin,
they need to know how much carbon dioxide to inject into
the bottles.

☛   **Your task**   From the two bottles of the "Real Thing" provided, find out
how much gas is needed per litre.

Based on a suggestion by I. Carpenter.

THE ROYAL
SOCIETY OF
CHEMISTRY

THE ROYAL
SOCIETY OF
CHEMISTRY

Age	14–18 years.
Time	120 minutes.
Group size	3.
Equipment & materials	Eye protection.

Items from the 'junk' list, for example, empty plastic lemonade bottles, (see page 20).

General
Test tubes, bungs, gas syringes, washing up bowls, conical flasks & bungs to fit them with a glass through tube (or side-arm flasks or side-arm boiling tubes or bung to fit the soft drink bottle with a glass through tube), plastic tubing, measuring cylinders, spatulas. Bunsen burners, tripods, gauzes, heatproof mats, clampstands. Balances. Simple tools (craft knife, pliers).

Selection of solids: salt, sugar, sand, anti-bumping granules. Access to water.

Per group
two small screw-top bottles of Coca Cola (250 cm^3 size).

Safety notes	See page 11.
Curriculum links	Gases dissolving in water, solubilities of gases with temperature, adding a solid releases the gas.
Possible approaches	Problem-solving at all levels of ability depending on students' appreciation of the complexity of the problem, *eg* just collecting the gas ... to ... realising the volume of water used is critical. (Volume of gas recovered is dependent on amount of water used and saturation of water with CO_2 – see There's Bags More Fizz in Fanta experiment.)
Evaluation of solution	The winners are the group producing the volume nearest to the correct value. (The judges will need to consider how best to determine the correct value for the volume of CO_2.)

Your notes

....

....

....

....

....

THE ROYAL
SOCIETY OF
CHEMISTRY

THE ROYAL
SOCIETY OF
CHEMISTRY

Atul only wanted half an apple, so his Granny put the other half in a cup of water. "Why did you do that, Granny?" asked Atul. "To stop the apple going brown, of course" replied Granny. "Is that the best way of stopping it going brown?" said Atul. "Let's find out, shall we?" said Granny. "We can try lots of different things from the kitchen."

☞ Your task

What do you think Atul and his Granny should do?

Based on a suggestion by P. Borrows.

THE ROYAL
SOCIETY OF
CHEMISTRY

Age	8–14 years.
Time	60 minutes.
Group size	2–4.
Equipment & materials	Eye protection.

General
Yoghurt pots. Aluminium foil – students may think light causes browning. Kettle.

Apples.

Salt, sugar, vinegar, lemon juice, bicarbonate of soda. Access to water.

Safety notes	See page 11.
Curriculum links	Chemical preservatives. (Biological oxidation.)
Possible approaches	Questions to ask students who need help are:- Do they need to use a whole apple each time, or can they use tiny slices? Would it be best to cut the apple up first, or to get everything else ready first? Is cold water better than warm water? Apart from water, what else might you safely try? (*eg* salty water, sugary water, vinegar, lemonade, bicarbonate of soda, anything else?) How can you make your tests fair?
	Each group could make a presentation to the class of their findings. At the end of each talk encourage 'members of the audience' to ask the speakers any questions, as one might do at any scientific meeting.
Extension work	Why is lemon or orange juice squeezed over a fruit salad containing apples, pears or bananas? Find out about the preservatives in your favourite food.

Your notes

....

....

....

....

....

....

....

....

THE ROYAL
SOCIETY OF
CHEMISTRY

THE ROYAL
SOCIETY OF
CHEMISTRY

The general thaw in East/West relations and swingeing government cut-backs in MI6 expenditure has resulted in people on cultural exchanges being used to carry sensitive diplomatic messages.

Development of new techniques using invisible ink have become essential as messengers using other techniques have mysteriously disappeared!

☛ Your task

Write a message in invisible ink on plain A4 paper. When dry the paper must be sent to MI6 together with instructions and materials for 'developing' the message. You are permitted 7 sheets of paper. You may send more than one message.

Based on a suggestion by J. Crellin.

THE ROYAL
SOCIETY OF
CHEMISTRY

Esso

THE ROYAL
SOCIETY OF
CHEMISTRY

Age	11–14 years. (All abilities).
Time	60 minutes.
Group size	2–4.
Equipment & materials	Eye protection.

Suggest students are warned in advance to bring an old shirt or a CDT apron to the session.

General
Cocktail sticks or empty old fountain pens or cotton swabs, beakers, test tubes. Pink paper, white paper.

Chemicals
Sulphuric acid, nitric acid, lemon juice, white vinegar, cobalt chloride, phenolphthalein indicator solution, sodium hydroxide. Potassium thiocyanate solution, iron(III)chloride solution.

Safety notes See Page 11. Care must be taken when developing those inks that require a heat source. (The ink should singe before the paper does.)

Curriculum links Chemical reactions.

Possible approaches Invisible inks consist of chemical solutions that are 'colourless' before developing, but which become visible when **(i)** heated, **(ii)** observed under ultra-violet light, or **(iii)** treated with other chemicals.

Invisible inks that respond to heat:-

1   Acids, *eg* sulphuric acid, nitric acid, lemon juice, white vinegar (colourless, go black with heat). Lemon juice is a very good magic ink. More concentrated acids are better than very dilute acids.

NB When hot acid is concentrated, it chars paper by reacting with the cellulose to produce black carbon.

$$\text{cellulose} \xrightarrow[\text{heat}]{\text{acid}} C(s) + H_2O(g)$$

2   Cobalt chloride (dilute cobalt chloride solution is light pink, almost colourless. If it is used as an ink the "invisible" writing goes blue on warming, due to dehydration of the salt. If you then breathe on the paper the writing once more disappears).

To develop the message, slowly pass the paper above a hot light bulb (check that the lamp is safely wired and earthed). Alternatively use a radiator or sunlight.

Invisible inks that require chemical treatment:-Phenolphthalein indicator solution (colourless, goes pink with alkali solution).

THE ROYAL
SOCIETY OF
CHEMISTRY

Potassium thiocyanate solution (colourless, goes red with iron(III)chloride solution). The Fe(III) ion reacts with the thiocyanate ion (SCN^-) to produce the red complex, $Fe(SCN)^{2+}$.

Evaluation of solution

Suggest credit is given for:-

1   The clarity (contrast and resistance to smudging) of the text after developing.

2   The extent to which the message can be detected before developing.

3   The wit of the message.

Extension work

To find other 'magic' inks.

Your notes

....

....

....

....

....

....

....

....

....

....

....

....

....

....

....

THE ROYAL
SOCIETY OF
CHEMISTRY

☛ **Your task**

Haynes Whiteners,
54 Henshaw House,
Paynes Parade,
St Pringle Bay.

Chief Chemist,
Science Laboratories,
Bunsen Road,
Test Tube Town.
CH4 7PH.

28th September

Dear Sir/Madam,

We are writing to ask your company of chemists to solve a problem
we have at the "Haynes Whiteners Factory". The factory makes 4
different white powders. Sometimes there is a mix up and the bags
of powders are not labelled. The factory needs to be able to tell
the difference between the white powders and identify which is
which. Your task is for your company to come up with some simple
tests and reactions which might show a difference between the 4
white powders.

We could arrange for your chemists to visit our very simple
laboratories to carry out some experiments. The equipment we have
available for your use include: beakers, Bunsen burners, test-tubes
and holders, Universal indicator and hydrochloric acid. Of course
this factory is very proud of its safety record and we would expect
you to work within the health and safety laws.

After the laboratory work, your chemists will be asked to explain
the tests to our factory workers and answer their questions.
Further, we would expect a report of your findings and
recommendations within the week.
Please do not hesitate to contact us if there is anything you are
unsure about.

Yours faithfully,

Based on a suggestion by S. Pringle.

THE ROYAL
SOCIETY OF
CHEMISTRY

THE ROYAL
SOCIETY OF
CHEMISTRY

Age	12–14 years. (All abilities, although more guidance will be needed for the lower abilities.)
Time	70 minutes.
Group size	3.
Equipment & materials	Eye protection.

General
Test-tubes (including pyrex ones), ignition tubes, test-tube racks, test-tube holders, beakers, glass droppers, glass stirring rods, spatulas. Microscope & slides. Bunsen burners, tripods, gauzes, heatproof mats, clampstands. Balances. Powerpack/leads/bulb. Electrolysis cell (as a distractor).

Universal indicator & indicator scales, litmus. Limewater.

Per group
White powders (solids should be powdered so that they are roughly of equal particle size). Choose 4 powders (approximately 10 g of each) from the list: magnesium oxide, sodium chloride, zinc oxide, ammonium chloride, sugar, citric acid, calcium carbonate, wax. (Students are told what the 4 powders are.) Hydrochloric acid (1 mol dm^{-3}) – 50 cm^3, sodium hydroxide (1 mol dm^{-3}) –50 cm^3.

Safety notes	See page 11.
Curriculum links	Physical and chemical changes, chemical reactions.
Possible approaches	Experiment is designed to make students think about how chemicals differ – physically and chemically. Tell students at start of lesson that they will need to ask permission if "unusual test" is required. Students are in competition with other companies for the business. Companies will be penalised for breaking health and safety laws, *eg* not wearing eye protection, untidy work.
Possible tests	*eg* Appearance, include under microscope. Adding water, adding universal indicator. Heating. Effects of adding acid/alkali. Weighing – density.

Conduction of electricity. Magnetic. (Distracters have been included.) ($CaCO_3$ gives off CO_2 when acid is added. $CaCO_3$ does not dissolve in water. Ammonium chloride sublimes on heating. Citric acid melts on heating to give an orange liquid. Ammonium chloride and citric acid turn indicator red.)

Suggested write-up	Students write a report for the factory boss. All the tests and results must be written up clearly with a conclusion of the easiest way(s) of identifying which powder is which. Also students will need a name for their company. The boss will be looking for an eye-catching and accurate presentation. Students might elect a representative from their group to present their findings – *eg* a POSTER – at end of the lesson.

THE ROYAL
SOCIETY OF
CHEMISTRY

☛ For an "unlabelled bags in the bakery" experiment:- use sodium hydrogencarbonate (sodium bicarbonate), sodium chloride, sugar and citric acid.

Your notes

....
....
....
....
....
....
....
....
....
....
....
....
....
....
....
....
....
....
....
....

THE ROYAL
SOCIETY OF
CHEMISTRY

☞ Your task

Your neighbour is in a hurry to go to the shops but has to
make a jelly. She asks you what is the quickest way to
dissolve the jelly cubes in water.

PLAN YOUR WORK. Try to make your experiment a fair test
of your ideas.

▲   The experiment must be repeatable.

Based on a suggestion by J.J. Palmer.

THE ROYAL
SOCIETY OF
CHEMISTRY

THE ROYAL
SOCIETY OF
CHEMISTRY

Age	11–15 years. ("The more able 12-13 year olds completed experiments successfully and came to satisfactory conclusions. A younger age group would need some parameters with which to work, as would less able students".)
Time	Total time = 140 minutes. But this time can be split into 2 lessons.
Group size	2–5 depending on ability and class size.
Equipment & materials	Eye protection.

General
You could make use of some of the items below if students were able to work in the Home Economics department (students could put equipment lists in for the technician):-

Heat source (Bunsen burner/hot tap/electric kettle/cooker/microwave). Containers (beaker/mixing bowl/pyrex dish/saucepan). Measurers (measuring cylinder/measuring jug). Stirrers(glass stirring rod/wooden spoon/fork). To make jelly pieces smaller (knife/scissors/cheese grater/potato masher/whisk/food processor). Thermometers. Stop clocks. Jelly moulds.

Per group
Jelly or a vegetarian alternative – 3 cubes. Water.

Safety notes	See page 11, safety aspects of eating jelly in laboratory.
Curriculum links	Rates of reaction. Dissolving.
Possible approaches	This problem is open-ended. Many scientific points to consider, *eg* temperature of solution, particle size of jelly, rate of stirring. If this experiment could be done in the Home Economics department students could actually eat their results (experiment has cross-curricular possibilities). Suggested write-up: student to write a note to neighbour telling her how to make a "quick jelly".
Extension work	☞ Ask students to "make the jelly solidify quickly". ☞ "How quickly can you get the jelly to dissolve without raising the temperature of the solution quickly"? (Try and explain why this worked.) ☞ FURTHER IDEAS:- Draw a plan of a machine that would make a jelly for a large party.

THE ROYAL
SOCIETY OF
CHEMISTRY

Your notes

THE ROYAL
SOCIETY OF
CHEMISTRY

Esso

☞ Your task

Mix two solutions of sodium thiosulphate and hydrochloric acid so that a colour change occurs after 1 minute.

Based on a suggestion by I. Taylor.

THE ROYAL
SOCIETY OF
CHEMISTRY

Age	12–16 years.
Time	One or two 90 minute periods depending on the interest shown.
Group size	2–3.
Equipment & materials	Eye protection.

General

Conical flasks, measuring cylinders, glass droppers, thermometers, stopclocks. Bunsen burners, tripods, gauzes, heatproof mats, clampstands. Plain paper with a cross drawn on it. (Graph paper – more able groups may be able to solve by extrapolation.)

Sodium thiosulphate solution (1 mol dm^{-3}), hydrochloric acid (1 mol dm^{-3}), access to water.

Safety notes	See page 11.
Curriculum links	Rates of reaction.
Possible approaches	In general students are asked to mix solutions of sodium thiosulphate and hydrochloric acid so that the colour change occurs in a set time, *eg* 1 minute. By changing concentrations, temperature, agitation *etc* the rate can be altered to that required. This experiment could be carried out after students have studied rates of reaction and what influences them, although with extra information it could be undertaken without previous knowledge.

Your notes

....

....

....

....

....

....

....

....

....

....

....

THE ROYAL
SOCIETY OF
CHEMISTRY

THE ROYAL
SOCIETY OF
CHEMISTRY

☞ **Your task**

> **To: Chemical Research Division**
>
> The crushing machinery in our quarry produces limestone chippings
> which are all the same size. These limestone chippings are used for a
> variety of purposes, one of which is to produce carbon dioxide gas by
> reacting the limestone chippings with dilute hydrochloric acid. This
> process is too slow for our needs*. As your task we should therefore
> be grateful if you could find out as accurately as possible how the
> speed of the reaction can be altered by using hydrochloric acid of
> different concentrations.
>
> Please submit detailed plans for experiments to test how the
> concentration of the acid affects the rate of reaction. To help us
> estimate the time needed to complete this work, please indicate
> clearly how many experiments will be needed and what conditions will
> be used for each one.
>
> Also indicate how any solutions you will need should be prepared. We
> should like to see a diagram of the apparatus for one experiment and
> a list of any other items you will need which are not shown on the
> diagram. Your report should include details of the amounts of
> materials to be used in each experiment and how these will be
> measured.
>
> You should explain the results of your experiments as they will be
> used to show the relationship between the acid concentration and the
> rate of the chemical reaction.
>
> * We cannot afford to install machinery to crush the chippings to
> smaller sizes, or the energy needed to heat the reaction vessel to a
> higher temperature.

Based on a suggestion by L. Ryan.

THE ROYAL
SOCIETY OF
CHEMISTRY

THE ROYAL
SOCIETY OF
CHEMISTRY

Age	14–16 years.
Time	70 minutes.
Group size	2–3.
Equipment & materials	Eye protection.

General
Beakers, conical flasks, conical flasks and bungs to fit them with a glass through tube, rubber tubing, gas syringes, measuring cylinders, stopclocks, cotton wool. Balances (grams) to 2 decimal places. Graph paper.

Calcium carbonate (marble chips approximately all the same size). Dilute hydrochloric acid (4 mol dm^{-3}).

Safety notes	See page 11.
Curriculum links	Rates of reaction.
Possible approaches	It is important to tell students that the amounts of hydrochloric acid and water should be measured as carefully as possible. The easiest way to compare the results from each part of the experiment is to construct a graph. Some students may follow the course of the reaction by observing the change in mass of the reaction mixture as carbon dioxide is given off. Others may collect the CO_2 given off in a gas syringe. "This challenge would make a good 'design an experiment' assessment for GCSE". Parameters such as stirring also need to be considered. (Used to supplement Salters' Chemistry Course – Minerals module.)

Suggested write-up

Chemical Research Division, Standish Chemical Company

Project Planning Report
Report submitted by: .... (name)
.... (date)

Diagram of Apparatus (label this to show the name of each item and show where chemicals are placed or products are collected).

List of other items required

Method

Extension work

Examine effect of **(i)** particle size and **(ii)** temperature on rate of CO_2 production – **NB** requires modification of student sheet.

Get students to draw a diagram of the plant, *eg* crushing machine, rollers *etc*. Is the process continuous or batch? Other questions, *eg* waste disposal, can also be considered.

THE ROYAL
SOCIETY OF
CHEMISTRY

Your notes

A Professor has invented a 'portable chemical circuit breaker' which when activated breaks an electrical circuit after a time delay of two minutes.

However, she is so absent-minded and forgetful that she can't remember **(i)** where she has put her laboratory notebook; **(ii)** how to make the 'circuit breaker' again and **(iii)** what it would be used for? However, she does remember that it contained a metal and dilute sulphuric acid. Also she can remember thinking to herself that it would enable her to become a millionaire!

☞ **Your task**

Help the Professor reconstruct her device (remember it must break the circuit two minutes after being activated) and suggest what it might be used for?

▲ Your device must be ready for operation at the end of the practical and will be activated when the Professor says so. Once activated the circuit should then be broken without you having to touch your device anymore. Evidence of circuit break is to be clearly visible.

NB A chemical reaction must be involved

▲ A diagram of your invention, with some explanation, will be needed.

Based on a suggestion by R.F. Kempa.

THE ROYAL
SOCIETY OF
CHEMISTRY

THE ROYAL
SOCIETY OF
CHEMISTRY

Age	14 years upwards.
Time	170 minutes (this time can be split into a 30 minute planning lesson where students have all the apparatus available to look at, and two 70 minute practical sessions).
Group size	2–3.
Equipment & materials	Eye protection. Items from the 'junk' list (see page 20) – to encourage creativity.

General
Filter paper, rubber and glass tubing of various sizes, small pieces of polythene sheeting, plastic syringes, stopclocks. Small electric bulb, electric battery (dry cell) suitable for use with light bulb.

Per group
Copper wire SWG 28 .... 35 cm
zinc foil.. 1.5 g
magnesium ribbon .... 2 g
sulphuric acid (1 mol dm⁻³)....................................... 300 cm³

copper(II)sulphate as "catalyst" (to enable zinc to be used). Depending on the knowledge of students it might be better to treat the zinc with dilute copper sulphate solution first.

Safety notes	See page 11. Students should be told not to concentrate the acid provided. Teachers need to be vigilant that groups do not generate gas in a sealed container (other than a syringe, plastic bag or similar) because of the risk of a pressure build up/spraying acid around.
Curriculum links	Reaction of acids with metal to produce hydrogen.
Possible approaches	There is no obvious best method. Approaches already noted:

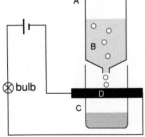

1   Reservoir of sulphuric acid dripping onto a magnesium fuse calibrated so that the magnesium dissolves after required time. Varying accuracy depending on construction.

2   Magnesium/acid reaction used to generate hydrogen. Hydrogen inflated balloon and movement of balloon breaks circuit contact. Variations on this method were most successful.

3   Magnesium/acid reaction used to generate hydrogen. Hydrogen production moves syringe which operates circuit breaker. Combination of acid concentration and distance moved by plunger used to vary time. Times variable depending on neatness of construction. All those adopting this method were short of time.

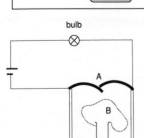

Top diagram: A = Top of plastic washing up bottle. B = Acid reservoir. C = Bottom of plastic washing up bottle. D = Magnesium ribbon contacts.
Bottom diagram: A = Magnesium ribbon contacts. B = Balloon. C = Acid reservoir. D = Magnesium ribbon. E = Glass tube.

Evaluation of solution   These are suggestions only:

1   Target time = 2 minutes. (**NB** Times which are too short approach the natural reaction time of magnesium with acid in a school laboratory, and therefore limit the problem. Times which are too long make the testing of devices a rather tedious procedure, rather than a high spot.)

2   Attainment of target time for circuit to be broken – subtract credit for each second off the target time.

3   Portability of fuse – credit could be awarded by subjective assessment of device constructed, taking into account fragility of device, ease of construction on site *etc.*

4   Size (smallness) of fuse – credit could be awarded by subjective assessment (as with (**2**) above).

5   The device should allow the "fuse" to be started and the circuit should be broken without any further intervention by the students.

Evidence of circuit break is to be clearly visible.

Extension work   Some students may wish to refine their device in their own time.

Your notes

....

....

....

....

....

....

....

....

....

....

....

....

THE ROYAL
SOCIETY OF
CHEMISTRY

☞   **Your task**

Pogueium (Pg) is a silver-blue metal which is found in nature
as the ore magowanite. Chemical analysis of magowanite
shows it to be pogueium chloride $PgCl_2$, a white, soluble
crystalline solid of melting point 415 $^{\circ}C$. Pogueium chloride
solution is an electrolyte. Pogueium reacts violently with acid
and effervesces gently with cold water. When calcium is
heated with pogueium oxide, calcium oxide is formed.
Magnesium is produced when pogueium is added to
magnesium sulphate solution. Deduce a method by which
pogueium metal may be extracted from magowanite.

▲   Draw a diagram of your invention with some explanation.

Based on a suggestion by S. Moore.

THE ROYAL
SOCIETY OF
CHEMISTRY

....

....

....

....

....

....

....

....

....

....

....

....

....

....

....

....

....

....

....

....

....

....

THE ROYAL
SOCIETY OF
CHEMISTRY

Age	15 years upwards.
Time	20–30 minutes.
Group size	1–2.
Equipment & materials	Pen, pencil, paper.
Curriculum links	Periodic table, extracting metals from their ores, reactivity series.
Possible approaches	Useful revision of reactivity series.
Notes	In competition reactions, the more reactive metal always forms a compound while the less reactive metal always ends up as the element.

$$PgO + Ca \longrightarrow CaO + Pg$$

$$Pg + MgSO_4 \longrightarrow PgSO_4 + Mg$$

Metals towards the top of the reactivity series are extracted by the electrolysis of their molten compounds.

Your notes

....
....
....
....
....
....
....
....
....
....
....
....
....
....

THE ROYAL
SOCIETY OF
CHEMISTRY

THE ROYAL
SOCIETY OF
CHEMISTRY

Mrs Johal has a problem – her class! They're very messy. Last week she had 3 jars – one with sawdust in, one with sand and a third one with salt. But now they're all mixed up – there is sand in the sawdust, there's salt in the sand and there's one jar with sand, sawdust and salt, all mixed up.

☞ **Your task**

How can Mrs Johal's class separate them?

Based on a suggestion by P. Borrows.

THE ROYAL
SOCIETY OF
CHEMISTRY

Age	7–11 years.
Time	60 minutes.
Group size	2-4.
Equipment & materials	Eye protection.

Items from the 'junk' list (see page 20) – to encourage creativity.

General
Filter funnels, filter papers, hand lens, tweezers, sieve.

Per group
You will need to provide a mixture of sand/salt/sawdust – say about a yoghurt pot half full – for each group. For younger students, or those less experienced in investigations, it may be better just to do the sand/sawdust mixture or the sand/salt mixture. Access to water.

Safety notes	See page 11.
Curriculum links	Dissolving, evaporating, filtration, floaters/sinkers.
Possible approaches	Questions to ask students who need help are:- Do you think magnets should work? What about a sieve? A home-made sieve? Could you use tweezers and a hand lens to pick out the pieces? What about using water? – Floaters/sinkers, dissolving.

☛ Few students realise that only a small amount of water is needed to dissolve the salt. If a large amount of water is used it can take a long time to evaporate. The final drying of the salt and the sand can be done in an oven.

Extension work	Students could design a large scale separation plant that works continuously. Separate chocolate bits from chocolate chip cookies.

Your notes

....

....

....

....

....

....

....

THE ROYAL
SOCIETY OF
CHEMISTRY

THE ROYAL
SOCIETY OF
CHEMISTRY

Kin Yu's mum has high blood pressure. Her doctor has told
her to cut down on the amount of salt she eats.

☞ **Your task**

Which sort of crisps would be best for her?

Based on a suggestion by I. Carpenter.

THE ROYAL
SOCIETY OF
CHEMISTRY

....

....

....

....

....

....

....

....

....

....

....

....

....

....

....

....

....

....

....

....

....

....

....

....

THE ROYAL
SOCIETY OF
CHEMISTRY

Age	7–16 years. (All abilities. Could be used for younger students as a 'fair test' exercise.)
Time	120 minutes.
Group size	2–4.
Equipment & materials	Eye protection.

General
For 11–16 years: filter funnels, filter papers, glass beakers, wash bottles, glass stirring rods, evaporating dishes. Bunsen burners, tripods, gauzes, heatproof mats, clampstands. Access to balances. Writing and/or display materials (sugar paper, felt tip pens *etc*) are also required. Binocular microscope (x 50 magnification).
Access to water.

Per group
1 packet of each of 3 different brands of ready salted crisps.

Safety notes	See page 11.
Curriculum links	Dissolving, filtration, evaporation. Healthy eating.
Possible approaches	Most 7–11 year olds will probably go for a taste test. Questions to ask students who need further help:- How can they make it fair? Does it matter if your mouth is already salty? Does it matter if the crisps are different sizes? Does it have to be a taste test? Could they wash the salt off and compare it? (One unexpected method adopted by a number of students in the 7–11 age range was to ask for a binocular microscope (x 50 magnification) and count the salt grains!)

11-16 year old students tended to wash the salt off the crisps with cold water (messy! other things dissolve as well) and collected the dissolved salt. Students could then evaporate the salt solution to dryness and weigh the salt residue, or use a hydrometer to measure how salty the water is. Alternatively, conductivity may be used.

Suggested write-up	Students to write a letter telling Kin Yu's mum which sorts of crisps would be best for her and how they arrived at their answer. Also a suitable poster could be produced (class effort) that might be put on display in the local health centre. Or students might produce a 'Health Education' video.
Evaluation of solution	Each group should put the crisps in order of decreasing saltiness. Older students should tackle the experiment more quantitatively.

Credit could be awarded for:

1   The correct order.

2   The quality of write-up.

3   The elegance of the approach.

THE ROYAL
SOCIETY OF
CHEMISTRY

Extension work   Students could find out what other foods should not be eaten if you have 'high' blood pressure. Students could also test for presence of starch in crisps. Comparison of salted peanuts.

Your notes

....

....

....

....

....

....

....

....

....

....

....

....

....

....

....

....

....

....

....

....

THE ROYAL
SOCIETY OF
CHEMISTRY

Your school has been on a day trip to France. On the way
home a smuggler slipped some crushed 'rough' diamonds into
a packet of salt which one of the girls was bringing back from
Calais.

Custom officers had been watching the smuggler for some
time and stopped the school party as they re-entered the UK
They now want to recover the diamonds and give you back
your salt... can you help them?

You know the diamonds and salt have a total mass of 5 g.

☞ **Your task**

Devise an experiment which will recover ALL the diamonds
and ALL the salt.

▲   Describe how you would check that ALL the salt had been
recovered.

Based on a suggestion by Tolworth Girls' School.

THE ROYAL
SOCIETY OF
CHEMISTRY

THE ROYAL
SOCIETY OF
CHEMISTRY

Age	11–16 years. (All abilities.)
Time	70 minutes.
Group size	2–3.
Equipment & materials	Eye protection.

General

Filter funnels, filter papers, glass beakers (100, 250 cm^3), conical flasks (100 cm^3 maximum), glass stirring rods, evaporating dishes, Bunsen burners, tripods, gauzes, heatproof mats, clampstands, balances should be available. Distillation apparatus (as a distractor).

Per group

Specimen labelled "CUSTOM SEIZURE":- A mixture of salt 'crystals' and 'rough' diamonds (*eg* anti-bumping granules). Total mass of mixture per group = 5 g (proportions of salt 'crystals' and anti-bumping granules must be noted by teacher). If you use this practical as an assessment it is easier if each group is given a mixture of the same composition.

Safety notes	See page 11.
Curriculum links	Dissolving, filtration, evaporation.
Possible approaches	Experiment could be used initially as an assessed practical (*ie* for design purposes where students have to suggest own methods). A couple of lessons later students could carry out their experiment and then present their results for assessment.

Few students realise that only a small amount of water is needed to dissolve the salt. If a large amount of water is used it can take a long time to evaporate. To stop the salt 'spitting' (leads to a loss in mass), only a small blue flame is required towards end of heating. A water bath could be used for more 'controllable' heating. Some students may struggle with the maths.

(This experiment could fit into a Forensic Science module.)

Suggested write-up	Personalise the student sheet for your school. Student to write a newspaper article about the event.

Your notes

....

....

....

....

THE ROYAL
SOCIETY OF
CHEMISTRY

THE ROYAL
SOCIETY OF
CHEMISTRY

Your aircraft has just crashed in the desert, and the last cupful of water has been spilled on the dry sand. You immediately scoop up the wet sand and put it into a plastic bag.

☛ **Your task**

How can you get that vital water back? You have just 90 minutes before you die from dehydration....

Based on a suggestion by I. Carpenter.

THE ROYAL
SOCIETY OF
CHEMISTRY

Age	12–13 years. (All abilities.)
Time	90 minutes.
Group size	2–3.
Equipment & materials	Eye protection.

General
The following items are available from the wrecked aircraft, or its occupants: plastic bags, yoghurt pots, aluminium foil, aluminium foil trays, beer cans, plastic lemonade bottles, rubber bands, wire, string, old pair of tights, blocks of expanded polystyrene or foam rubber. Pliers, craft knives.

Furthermore, the desert environment provides unlimited quantities of sand, and a source of radiant energy – a lamp with a 60 W bulb should be provided to simulate the desert sun (check that it is safely wired and earthed). A stage floodlight would be exciting if available!

Per group
A plastic bag containing 100 g sand mixed with 25 cm^3 water, *ie* wet sand.

Safety notes	See page 11. Warning about mains electricity and water.
Curriculum links	Evaporation, condensation.
Possible approaches	Fits in well with classwork on evaporation and condensation. Some help on cooling steam may be necessary.

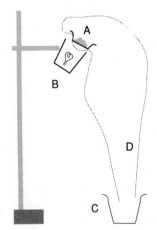

☛ A wide, clear plastic tube could be provided, so that the principle of extraction of water using centrifugal force could be explored, *cf* spin dryer.

☛ Plastic dish covered in foil, inside dish, porous pot (sealed) containing wet sand. Bottom of dish insulated. At the top a lamp is shone on the dish.

A = Wet sand in aluminium dish
B = 60w bulb
C = Yoghurt pot
D = Plastic bag set at an angle

Suggested write-up

Taped interviews conducted with the 'survivors' of the aircrash on their return to the UK by science reporters. The novel water 'extractors' already making headline news need the students' personalised explanations.

THE ROYAL
SOCIETY OF
CHEMISTRY

Evaluation of solution   These are suggestions only:

1   Just before they are about to die, the 'survivors' (groups) present the judges with the water they have freed from the sand.

2   The judges will measure the volume of water collected. The winning group is that with the largest volume of water.

3   In the event of a tie, the group with the most appetising sample of water will be declared the winners.

Extension work   Cross-curricular possibilities with the geography department – *ie* water supply in desert regions.

Your notes

....

....

....

....

....

....

....

....

....

....

....

....

....

....

....

....

....

THE ROYAL
SOCIETY OF
CHEMISTRY

☞ Your .task

You are given a collection of plant materials – leaves, petals, and so on. They are all coloured. Extract the pigments (colouring substances) from them, and decide how many different pigments you have got altogether.

▲ **HINTS:** Remember that some plants may contain more than one pigment mixed together. Remember, too, that the same pigment may be present in more that one plant, so don't count it twice ... as long as you are sure it really is the same pigment.

▲ Display your results in a way which will allow people to see how you arrived at your answer.

Based on a suggestion by P. Borrows.

THE ROYAL
SOCIETY OF
CHEMISTRY

Age	13 years upwards. (Students need a prior knowledge of chromatography.)
Time	120 minutes.
Group size	2–4.
Equipment & materials	Eye protection.

General

A range of apparatus for the commonly used paper chromatographic techniques: stoppered boiling tube into which filter paper strip can be inserted, beakers, petri dishes, glass droppers, scissors, paper clips, pencil & ruler (measuring to mm).

Apparatus for extracting pigment from plant materials: pestle and mortar, filter funnel, flasks and beakers, test tubes. Possibly access to a centrifuge. Sand.

Access to water and ethanol and/or propanone. (**NB** It is suggested that distilled water should NOT be used – as this is often acidic, and some pigments are indicators, this may confuse the issue.)

Per group

Students are likely to want large amounts of filter paper, both for chromatography and filtration (a box per group).

The same plant materials, say 5 or 6 different types, including:-
(i)   red-coloured leaves (*eg* copper beech);

(ii)   green leaves (*eg* grass, spinach);

(iii) red cabbage or beetroot (not pickled);

(iv)   flower petals of at least 2 different colours (*eg* yellow and purple).

Safety notes	See page 11. Naked flames should not be allowed with the flammable solvents. Heating is probably not necessary, but hot water from a kettle could be made available for a water bath if requested.
Curriculum links	Chromatography.
Possible approaches	The judges will need to run a 'control'. With grass, it should be possible to separate chlorophyll and xanthophyll, but you are unlikely to get more than one yellow and one green band. With copper beech leaves it should be possible to detect a red and a green pigment – but is the green pigment the same as that in grass? 13–14 year olds are not likely to be familiar with Rf values *, but some may make qualitative judgements (*eg* by similarity of colour) based on 'fair test' comparisons.

$$*[Rf = \frac{distance\ travelled\ by\ substance}{distance\ travelled\ by\ solvent\ front}]$$

THE ROYAL
SOCIETY OF
CHEMISTRY

Evaluation of solution   Credit could be given for:

1   Evidence of a systematic approach.

2   Quality of display.

3   Each correctly identified pigment.

4   Some evidence of 'fair testing', controlling variables, Rf values or equivalent *etc*. DEDUCT credit for each wrong pigment, or if same pigment counted twice.

Extension work   Find out about synthetic dyes (1850s onwards).

Your notes

....

....

....

....

....

....

....

....

....

....

....

....

....

....

....

....

....

THE ROYAL
SOCIETY OF
CHEMISTRY

Esso

You are on a 'Survival' Course, your group has been left in the 'wilds' for a few days with only a limited amount of water and food. It is a scorching hot summer's day. Early evening you discover that the person carrying the water has drunk it all and there is only powdered potatoes or packet soup to eat. Clearly, in order to eat and drink tonight you must obtain some water. Luckily you come across a muddy pond next to a rubbish tip. Fortunately, you also notice old bits of laboratory equipment in the rubbish tip. There is a wood fire beside the pond with charcoal in it.

☞ **Your task**

Obtain some pure water in order to prepare a meal and to have water to drink the following day. You must also find a way of showing that the water is pure.

Based on a suggestion by R.F. Kempa/K. Davies.

THE ROYAL
SOCIETY OF
CHEMISTRY

THE ROYAL
SOCIETY OF
CHEMISTRY

Age	12–14 years. (Some previous experience of filtration, distillation and use of glassware is needed.)
Time	100 minutes (this time can be split into a 30 minute planning lesson where students have all the apparatus available to look at, and a 70 minute practical session).
Group size	2–4.
Equipment & materials	Eye protection.

General

Glass beakers (100, 250, 400 cm^3), large plastic trays, test tubes, boiling tubes, test tube holders, conical flasks & bungs to fit them with a glass through tube (or side-arm flasks or side-arm boiling tubes), straight & bent glass tubing, rubber tubing (5 cm & 30 cm lengths), funnels, filter papers, plastic sieves, glass droppers, evaporating basins, thermometers, paper towels. Bunsen burners, tripods, gauzes, heatproof mats, clampstands.

Charcoal.

Per group

'WATER SAMPLE FROM MUDDY POND':- tap water (400 cm^3) mixed with soil (about half a handful), green ink to colour noticeably + salt (4 g) mixed.

Safety notes	See page 11. Check students' apparatus before they start heating. Remind students that if their water boils too vigorously they are to turn down the flame. Small blue flame only required.
Curriculum links	Filtration, distillation, temperature.
Possible approaches	Fits in well with classwork on evaporation and condensation. Some help on cooling steam may be necessary. Pond contents could be varied, for example oil could be added to the water.
Suggested write-up	Student to write a diary entry for day 1 of the Survival Course (as part of the 'Survival Course' assessment procedure).
Evaluation of solution	These are suggestions only:

1 Firstly judge against students' criteria for success.

2 That some purification was achieved.

3 Does method take into account soluble impurities as well as insoluble impurities?

4 Some tests done on water (visual + boiling point + others).

5 Distillation.

Extension work	Students find out what is in pond water.

THE ROYAL
SOCIETY OF
CHEMISTRY

Your notes

THE ROYAL
SOCIETY OF
CHEMISTRY

☛ **Your task**

MEMO ·

To: Form 2N
From: The Headteacher

I suspect that the school secretary is trying to
do me in by putting glass shavings in my sugar
bowl. Mr Roberts tells me that you are expert
scientists, so would you please analyze the
enclosed sample and report back to me.

Headteacher ·

P.S. I would like the sugar back!

Fifteen minutes before the end of this laboratory session a
meeting will be held to discuss the methods you have used
and your findings and conclusions! IS THE SECRETARY
GUILTY? All project scientists are required to attend. Group
reports will be forwarded to the Headteacher. Be prepared to
argue your case!

Based on an idea by the SSCR in Wiltshire.

THE ROYAL
SOCIETY OF
CHEMISTRY

THE ROYAL
SOCIETY OF
CHEMISTRY

Age	12–15 years. (A revision exercise for 15 year olds.)
Time	90 minutes for 12 year olds (this time can be split into a 30 minute planning lesson and a 1 hour practical session). 70 minutes for 15 year olds (including written report).
Group size	2–3.
Equipment & materials:	Eye protection.

General
Filter funnels, filter papers, glass beakers, conical flasks, glass stirring rods, evaporating dishes. Bunsen burners, tripods, gauzes, heatproof mats, clampstands. Distillation apparatus (as a distractor). Microscope.

☞ Teacher demonstration: Apparatus for vacuum evaporation (see SCSST & ASE booklet, "Experimenting with Industry No.4 SUGAR CHALLENGE", p4, Fig 3).

Per group
Specimen labelled "SAMPLE FROM THE HEAD'S SUGAR BOWL" – a mixture of sugar and 'crushed glass' (**NB** the glass and sugar should be of equal particle size, students should not be able to distinguish the glass pieces from the sugar). However, we strongly recommend an alternative to crushed glass, *eg* anti-bumping granules.

Safety notes	See page 11. If glass is used it is essential to warn students of dangers of glass shavings, and to dispose of unused 'glass and sugar' mixture carefully at end of practical session. However, we strongly recommend that crushed glass is not used but an alternative, *eg* anti-bumping granules.
Curriculum links	Dissolving, filtration, evaporation, chemical changes, crystallisation.
Possible approaches	The first part of the experiment, to find out if there are 'glass shavings' in the sugar bowl, is straight-forward. The second part of the experiment, getting the sugar back is more difficult:-

☞ "When given to a top-set 4th year, many students managed to caramelise the sugar because they used a naked flame". (**NB** Getting rid of the water cannot be done just by heating because at high temperatures the sugar starts to change chemically; remember that as the concentration of a solution increases, so does its boiling point.)

☞ Some students may concentrate the sugar solution using a water-bath and then leave this solution to evaporate naturally (one group who did this found that evaporation of their liquid took a couple of weeks). Sugar is very soluble in water and will not crystallise unless most of the water is removed.

☞ Few students realise that only a small amount of water is needed to dissolve the sugar initially. If a large amount of water is used it can take a long time to evaporate.

THE ROYAL
SOCIETY OF
CHEMISTRY

A useful discussion may follow on how to get rid of the water 'quickly' without also chemically changing the sugar. The teacher may wish to demonstrate to the class how to evaporate the sugar solution at a lower temperature with the help of a tap vacuum.

Suggested write-up

Personalise the student sheet for your school and your class. Students will often write letters to the Headteacher who may even respond to their creative work. (It would also be courteous to let the school secretary know what you are doing!)

Extension work

Find out how sugar is obtained from sugar beet. For teachers' information:-

"Chemistry in Action!" - A Publication for Chemistry Teachers in Ireland. (Issue 20, Autumn 1986, "SUGAR".)

Your notes

....

....

....

....

....

....

....

....

....

....

....

....

....

....

THE ROYAL
SOCIETY OF
CHEMISTRY

An old iron cargo ship, carrying a cargo of ammonium chloride was wrecked on a sandy beach off the chalky white cliffs of Dover. The cargo was saturated with salty sea water, but subsequently dried out.

Pounded by recent storms, this mixture of 5 substances has been reduced to the sample given to you.

Local environmental groups have demanded that these five compounds should be separated and either recycled or returned to their original location.

☞ **Your task**

Separate this mixture to give a pure sample of each component of maximum yield.

▲ A sample of each component in its pure form is available if you wish to carry out any preliminary investigations.

Based on a suggestion by J. Crellin/V. Herbert.

THE ROYAL
SOCIETY OF
CHEMISTRY

Age	16–18 years.
Time	120 minutes.
Group size	2–3.
Equipment & materials	Eye protection.

General
Magnets, crucibles, evaporating basins, test tubes, small pyrex, or hard glass test tubes, test tube holders, filter funnels and papers, glass beakers (100 cm³ maximum), glass stirring rods, spatulas, rocksil wool. Bunsen burners, tripods, gauzes, heatproof mats, clampstands. Top pan balances.

Per group
Sodium chloride, ammonium chloride, silver sand, iron filings, calcium carbonate and a mixture containing 25 g of each of the above chemicals.

Safety notes
See page 11.

Curriculum links
Sublimation, magnetism, filtration, evaporation, solubility of salts in water. (Groups 1 and 2: chemical elements.)

Possible approaches
Each group should be provided with a sample of each component in its pure form, for any preliminary experiments they may wish to do. Before adding water it is best to separate **(i)** the iron filings from the mixture with a magnet, and **(ii)** the ammonium chloride by sublimation. Then add water to the remaining sand, calcium carbonate and sodium chloride mixture and filter. Filtrate of sodium chloride solution is evaporated. Calcium carbonate and sand are left on the filter paper. Addition of hydrochloric acid to the sand/calcium carbonate mixture forms a solution of calcium chloride (sand removed by filtration). Addition of sodium carbonate solution precipitates calcium carbonate (by double decomposition). Calcium carbonate and sand mixture may also be separated by allowing the two components to settle out over a period of several weeks.

☛ Cargo can be anything, *eg* organic material, coloured material becomes a possible extension.

☛ Few students realise that only a small amount of water is needed to dissolve substances. If a large amount of water is used it can take a long time to evaporate (**NB** keep large beakers out of reach of students).

Suggested write-up
Students to write a letter to local environment group with results.

Evaluation of solution
These are suggestions only:

1   The yield of each component should be recorded.

2   The separated components should be assessed for acceptable purity.

3   The winners are the group that produces the greatest total mass of separated components of acceptable purity.

Your notes

....
....
....
....
....
....
....
....
....
....
....
....
....
....
....
....
....
....
....
....
....

THE ROYAL
SOCIETY OF
CHEMISTRY

☞ Your task

Using the materials and equipment provided make:

1   The longest lasting bubble.

2   The biggest single bubble.

3   Lots of small bubbles.

Safety

Do NOT blow bubbles into eyes (it stings!).

▲   A wet, soapy slippy floor is dangerous so mop up any spills
and watch where you put your feet!

▲   Do NOT put bubble mixture in your mouth!

Based on a suggestion by I. Carpenter/R. Lewin.

THE ROYAL
SOCIETY OF
CHEMISTRY

Age	7–16 years.
Time	120 minutes.
Group size	3–6.
Equipment & materials	Eye protection.

General

Length of thickish wire – 1 metre (or various sizes of wire hoops), pliers, fly-swat. Filter funnels, yoghurt pots, plastic lemonade bottles, pipe cleaners, small syringes, glass droppers, measuring jug, bowls, scissors, stopclocks, 30 cm rulers.

4 or 5 different brands of washing up liquid, kitchen soap, glycerine, sugar, salt and water. (**NB** For younger students the problem can be simplified by giving no choice of detergent, and no glycerine. For all but the oldest the glycerine is best avoided, or there will be too many variables.)

Safety notes

See page 11. In addition see safety notes on student sheet.

Curriculum links

Surface tension.

Possible approaches

"Good experiment for a hot sunny day outside!". Lots of variables to consider. Agreement must be reached on how to measure bubble size – in one school 7-8 year old students decided to catch the bubble on a hoop and hold it near a ruler. (You might also have a session on bubble blowing without recording.)

For an article that describes the preparation and use of a bubble chamber, in which students can view a bubble from the inside out, see Chemistry for kids: "Invitation to Chemistry through a Large Soap Bubble Chamber" Sanae Sato, J.Chem.Educ., July 1988, Vol 65, p616.

Evaluation of solution

Each group could, when required to do so by the judges, use their optimum solution to blow three bubbles. Any bubble surviving for at least 10 seconds could be measured. In addition each group could present a visual display showing how they arrived at the optimum solution.

Judges should look for an understanding of the need to control variables and a systematic approach to the problem. Marks for the display could be added to those of bubble size to identify a "winning group".

Record sheet

(for 7–10 year olds)

How can we best measure the biggest bubble?

Ideas from ....group:
....
....

THE ROYAL
SOCIETY OF
CHEMISTRY

Experiment 1

Who blew the biggest bubble in .... group.

Mark the winner with a red *
Name:
1 ...... 2 ...... 3 ...... 4 ...... 5 ...... 6 ....

Piece of equipment best for making biggest bubbles was:
large wire hoop   medium wire hoop   small wire hoop
plastic bottle top   yoghurt tub pot
Any other? -

Experiment 2

Who blew the longest-lasting bubble?

Mark the winner with a green *
Name:
1 ...... 2 ...... 3 ...... 4 ...... 5 ...... 6 ......

Time bubble lasted:
1 ...... 2 ...... 3 ...... 4 ...... 5 .... 6 ......
Best mixture for blowing bubbles contained
.... cm^3 washing up liquid + ... cm^3 water. How
did your group arrive at these quantities?

....
....

Experiment 3

Make a list of the pieces of apparatus you used to see
which was best for blowing lots of small bubbles. Mark
the one your group found best using a blue *
1.
2.
3.
4.
5.
6.

Extension work

Invent a bubble-making machine so there is no need to blow one
bubble at a time using your 'lung power'.

Your notes

....

....

....

....

THE ROYAL
SOCIETY OF
CHEMISTRY

☛ **Your task**

Build a boat powered by soap to carry a ten gram mass and to
see how many times it will go round a clampstand.

▲   **HINT:** You will need to change the water each time you try.

Based on a suggestion by S. Lindley.

THE ROYAL
SOCIETY OF
CHEMISTRY

THE ROYAL
SOCIETY OF
CHEMISTRY

Age	11–13 years. (All abilities.)
Time	Anything up to an hour depending on how many "attempts" students have at the circuit.
Group size	2–3.
Equipment & materials	Eye protection.

Per group
A plastic washing up bowl, a clampstand (no fittings), cotton, aluminium foil, a small piece of soap, a 10 g mass, polystyrene, pins.

Safety notes	See page 11.
Curriculum links	Surface tension.
Possible approaches	The students make a boat from aluminium foil and polystyrene containing a 10 g mass. A small piece of soap is pinned to the stern (back) of the boat. The boat is attached to a clampstand in the middle of a plastic washing up bowl of water by a length of cotton. Students have to see how many circuits of the bowl the boat can make before it needs refuelling.

At least 30 laps of the bowl will be completed on one filling by an experienced "captain". Some students will do some major re-designing when they realise that in a washing up bowl the biggest boat is not the best! The students will need prompting to replace the water – rather than the soap – for another try. Count the 10 g masses before you start, they tend to get thrown away with the boats at the end of the lesson.

The boat is able to work because of surface tension. The soap pinned to the stern of the boat slowly dissolves. The surface tension of soap solution is less than that of water, therefore the 'pull' of the water in front of the boat is greater than the 'pull' of the soap solution. The boat moves forward.

Extension work

Investigate whether the type of soap makes any difference. Would the boat work in a swimming pool – where chlorine is in the water?

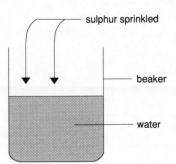

Excellent demo or "magic trick". Sprinkle flowers of sulphur onto water. When you touch the surface with a soapy finger the sulphur immediately falls like snow to the bottom of the beaker. VERY DRAMATIC. Set homework for students to try out herbs, flowers *etc.*

THE ROYAL
SOCIETY OF
CHEMISTRY

Your notes

THE ROYAL
SOCIETY OF
CHEMISTRY

☞ **Your task**

"Burning candles lose mass". Use this knowledge to make a
timer.

▲ Test your ideas

Based on a suggestion by S. Lindley.

THE ROYAL
SOCIETY OF
CHEMISTRY

THE ROYAL
SOCIETY OF
CHEMISTRY

Age	11–13 years. (The less able will need guidance.)
Time	60 minutes.
Group size	2–3.
Equipment & materials	Eye protection.

Per group

Pins, clampstand, cork, heatproof mat, card or stiff paper, wooden splint, plasticine or blu-tack to balance the splint, a timer, a 30 cm ruler, a pencil.

2 or 3 birthday cake candles (the candles with the spiral markings seem to work better than the plain candles).

Safety notes See page 11. If the candle is allowed to burn down too far it sets light to the splint. If the scale is too close to the flame it too may burn.

Curriculum links Balancing, friction.

Possible approaches Students make a timer by fixing a birthday cake candle on one end of a splint, pivoted with a pin, in a cork held in a clampstand. They will need to find a way of making a scale against which the other end of the splint moves. This scale is then calibrated using a timer. The core of the problem is to get the splint to pivot easily - but not too easily. It is possible to get repeatable results, given care.

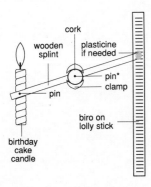

eg A birthday cake candle timer

☞ *the friction between the pin and the wooden splint is important. It can be altered by changing the amount of pressure of the pin against the cork.

Your notes

....
....
....
....
....

**THE ROYAL
SOCIETY OF
CHEMISTRY**

THE ROYAL
SOCIETY OF
CHEMISTRY

Carbon dioxide is widely used in fire extinguishers (coloured black to distinguish them from water-containing extinguishers which are red). Because carbon dioxide is heavier than air it smothers and extinguishes fires by preventing further oxygen from reaching the source of the flames.

However, one of the difficulties in putting out big oil fires in the open air is that the wind blows away the carbon dioxide gas, allowing oxygen to reach the fire, keeping it burning. Ideally, we should apply the carbon dioxide not as a gas but as a blanket of thick foam.

☛ **Your task**

Make as much foam as possible (measured in a very large container) using:

1   Any combination of the 3 liquids (you are allowed a maximum volume of 20 cm^3 of each).

2   Any combination of the 3 solids (in this case you are allowed a maximum of 6 spatula fulls of each solid).

These can be mixed in any order, but must not be shaken or stirred.

Based on a suggestion by I.M. Childs/M. Goodall.

THE ROYAL
SOCIETY OF
CHEMISTRY

Age	12–15 years. (All abilities. Could be used for younger students as a 'fair test' exercise.)
Time	70 minutes.
Group size	2–3.
Equipment & materials	Eye protection.

General

Large measuring cylinders (500, 1000 cm^3) or empty 1 litre plastic lemonade bottles with tops cut off, pestles and mortars, spatulas. Sodium hydrogencarbonate, sulphuric acid (2 mol dm^{-3}), aluminium sulphate, washing up liquid, washing powder, water, food colouring – optional (see possible approaches below).

Safety notes	See page 11.
Curriculum links	Colloids. Combustion and Firefighting. Carbon dioxide.
Possible approaches	A foam is a colloidal system in which a gas is dispersed in a liquid. Construction of a table to record results would be useful and aid systematic working. The foam is formed when a sodium hydrogencarbonate solution is mixed with a solution containing a detergent and aluminium sulphate (or any weak acid). Carbon dioxide gas is produced which is trapped by the detergent.

Trialled with a group of mixed ability second years, students were not given washing powder, just liquid detergent, to reduce variables. 500 cm^3 measuring cylinders were fine for most students, but one or two needed to use the litre measuring cylinders. (It might be possible to use 500 cm^3 beakers but they would really be too wide.) The foam makes it very difficult to read the calibrations on the measuring cylinder – colouring the mixture might help. Students were good at keeping a record of the amounts they used – which has to be quoted for their 'best' result to count. The greatest volume of foam was 800 cm^3.

☞ **NB** Conkers make wonderful foam! Boil conkers up with water: peel conkers, mash them up, put in 250 cm^3 beaker a third full of water. Boil for 5/10 minutes. DECANT SOLUTION. To make the froth: add sodium hydrogencarbonate (approximately a dessert spoon) and aluminium sulphate (sufficient acidity).

☞ Frothing agent obtained is the same as that found in shaving foams and fire extinguishers (a C_{10} alcohol).

Extension work

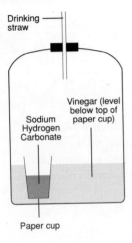

Drinking straw

Vinegar (level below top of paper cup)

Sodium Hydrogen Carbonate

Paper cup

Students could go on to design a 'foam launcher', and make a fire extinguisher that operates. They will need to consider the technical problem of delivering the foam to put out fire.

☞ Name some other examples of foams and their useful applications, *eg* shaving cream and whipped cream.

Your notes

....

....

....

....

....

....

....

....

....

....

....

....

....

....

Recipes for cola are closely guarded secrets, and soft drinks manufacturers are continually trying to find out as much as possible about their rivals cans of cola.

You have been asked by an independent laboratory to help them in their investigations.

☛ **Your task**

Find out as much as you can.

Based on a suggestion by P. Ward.

THE ROYAL
SOCIETY OF
CHEMISTRY

THE ROYAL
SOCIETY OF
CHEMISTRY

Age	Experiment can be adapted to suit any age group (see possible approaches below).
Time	Depends on investigation.
Group size	2–4.
Equipment & materials	Eye protection.

General
Assorted equipment, depending on approach.

Per group
One can of cola.

Safety notes	See page 11.
Curriculum links	Depends on investigation.
Possible approaches	Very open-ended. A selection of the following could constitute a useful investigation.

Experiments

1   Volume of cola in can.

2   Density of cola in can.

3   Volume of gas released on opening under water.

4   Tests on gas collected – combustion/limewater.

5   What is can made of? Effect of magnet, acid, alkali on metal. Tests for Al^{3+}. Density of the metal.

6   pH of the cola.

7   Measure of total acidity by titration (Decolorise with activated charcoal for some of the tests.)

8   Evaporate down – any solid residue?

9   Column chromatography to separate dyestuffs in the cola. Caramel?

10   Distil the cola – what is the distillate? Boiling point?

11   Is the cola optically active? What sugar is present? How much?

12   What is the can painted with – resistance to solvent attack?

(Use your imagination for anything else. Caffeine, benzoic acid? – read the label.)

Suggested write-up	Open day display could be produced by students.
Extension work	Other colas, lemonade or other cans of carbonated soft drinks – what differences arise? If students compare different soft drinks – panel discussion about why they prefer one to another!

THE ROYAL
SOCIETY OF
CHEMISTRY

Your notes

THE ROYAL
SOCIETY OF
CHEMISTRY

When a beautiful bouquet arrives from a good florist it is usually accompanied by a sachet of cut flower preservative. Since this really does seem to work, we have often wondered what was in it, or how we could buy it to use on our humbler shop purchases. Now the Guernsey Flowers Information Bureau has let us into the secret. According to them you can make your own preservative solution by dissolving half a tablespoonful of sugar to one teaspoonful of bleach in a pint of water.

Your local florist is interested in making their own preservative solution but doesn't have the time to investigate the claims made by the Guernsey Flowers Information Bureau. As a result they have approached your school for help?

☛ **Your task**

Is this claim valid?

Based on a suggestion by K. Davies.

THE ROYAL
SOCIETY OF
CHEMISTRY

THE ROYAL
SOCIETY OF
CHEMISTRY

Age	12–14 year olds. (All abilities.)
Time	70 minutes (students then need to compare the flowers every week, noting any differences).
Group size	2–3.
Equipment & materials	Eye protection.

General
Measuring cylinders, measuring jug, beakers, glass droppers, glass stirring rods, teaspoons, tablespoons.
Sugar, bleach, water.
Flowers
Carnations (standard and sprays)/roses/ freesias.

Safety notes	See page 11. It is advisable to pre-dilute the bleach solution.
Curriculum links	Chemical preservatives. (Biochemical decay.)
Possible approaches	Could be used as a 'fair test' exercise. Apparently the effect of the preservative solution is most marked on flowers such as carnations (standard and sprays), which have been kept for as long as 3 weeks, and roses. (Freesias have also lasted longer in the preservative solution.) Don't use chrysanthemums as they keep reasonably well anyway.

☛   Always keep flowers in a cool place.

Reason that preservative solution works:

1   Sugar is a food.

2   Bleach kills bacteria. (The flower stems have water channels. Bacteria clog up these channels – the bleach kills the bacteria.)

Suggested write-up	Students write a report for the florist.
Extension work	What sugar/bleach/water ratio is best? What is the effect of different bleach concentrations? (If this is investigated then clearly it will be necessary to make up different pre-diluted bleach solutions, including chlorine and non-chlorine bleaches.)

It has also been suggested that "Andrews" is good for flowers – carbon dioxide goes up the stem and stops them going limp. Students could test if this is true.

Your notes

....

....

....

THE ROYAL
SOCIETY OF
CHEMISTRY

THE ROYAL
SOCIETY OF
CHEMISTRY

An unlabelled drum of liquid has been washed up on a
nearby beach. The local police and fire brigade were called to
the scene and have obtained a sample of the liquid for
analysis.

☛ Your task

Find the density of the liquid and its solubility in water.

▲ You should compare your results with the details in the data
table and try to name the liquid.

▲ Check that your method is safe before carrying out your
experiment.

Based on a suggestion by Wallace Hall Academy.

THE ROYAL
SOCIETY OF
CHEMISTRY

Data sheet

Liquid	Density (g/cm^3)	Solubility in Water
Pentane	0.62	Insoluble
Hexane	0.66	Insoluble
Heptane	0.68	Insoluble
Cyclopentane	0.72	Insoluble
Cyclohexane	0.77	Insoluble
Pentene	0.64	Insoluble
Hexene	0.67	Insoluble
Ethanol	0.80	Soluble
Propanol	0.83	Soluble
Trichloromethane	1.49	Insoluble
1,1,1-Trichloroethane	1.33	Insoluble
Water	1.00	Soluble

Age	14–15 years. (The instructions are adequate for "top" groups, but students of lower ability may need the clues (see possible approaches) to set them on the right course.)
Time	25–30 minutes.
Group size	2–3.
Equipment & materials	Eye protection.

General
Selection of measuring cylinders (better than giving one size), test tubes, glass droppers, beakers (100 cm^3). Balances – (grams) to 2 decimal places.

Per group
Approximately 20 cm^3 of liquid. **NB** The liquid which the groups are given is an important issue – not all of the liquids are suitable/safe. It doesn't matter if they are listed in the table, as long as the one given to students is safe. Therefore, it is recommended that you use ethanol, propanol or water.

Most of the liquids should preferably be kept in the fume cupboard.

Safety notes	See page 11. Warn students that the liquid could be flammable and poisonous. No naked flames. **NB** Use ethanol, propanol or water.
Curriculum links	Density.
Possible approaches	It is not possible for 11 groups to explain fully their intentions to one judge before starting, you could therefore inform students that they may do any experiment not involving tasting, smelling or flames, so as to avoid queues.

☞ Only one balance can cause some delay to students.
☞ A demonstration to show how a more accurate measurement of density could be made using a density bottle. Pipettes/burettes, and even volumetric flasks could be used as equipment. (Accurate density measurement is limiting factor.) Errors should be discussed, especially related to the volume of liquid used.

NB Before each experiment judges should determine their own density values for the liquid they choose to use. Also, it is advisable to give an acceptable density range, rather than a single figure.

Clues	A. Density

1 Density is found by finding the mass of a liquid and dividing the value by its volume. density = mass/volume

2 Find the mass of the liquid in a beaker and measure its volume in a measuring cylinder.

B. Solubility

1   Solubility is found by adding a small amount of liquid to water and checking if it dissolves.

2   If the liquid is soluble, the 2 liquids appear as one. If the liquid is insoluble, 2 separate layers are seen.

Evaluation of solution   Credit could be awarded for:

1   Finding the density.

2   Finding the solubility.

3   Naming the compound.

4   Good write-up.

If a group seeks assistance, credit should be lost for every visit to the judges or 'discussion' with another group.

Extension work   Students could make a special floater (hydrometer) from a plastic drinking straw and plasticine to determine densities of liquids by comparison.

Your notes

....

....

....

....

....

....

....

....

....

....

....

....

You are the scientist working for a consumer magazine. This month they are investigating chlorine-containing bleaches; they want to find out which bleach is the best buy.
(**NB** Chlorine is the active chemical in bleach.)

☞ **Your task**

Determine how much available chlorine is present in four different bleaches, and which bleach is the best buy.

▲ Each bleach has already been diluted 50 times by water.

Based on a suggestion by S. Robilliard.

Age

16–18 years. A simplified version with younger students (14-15 year olds) has been used – they were given bleach solutions which had been pre-diluted in such a way that equal volumes of the different bleach solutions cost the same amount.

Time

90 minutes.

Group size

4.

Equipment & materials

Eye protection.

General

Burettes, funnels, volumetric flasks (available but not visible). Conical flasks and white tiles. Pipettes and safety fillers. Indicator paper (as a distractor).

Solutions of four different brands of chlorine-containing bleach (diluted 50 times by water). **NB** Environmental friendly bleaches do not contain chlorine.

Per group

Sodium thiosulphate solution (0.1 mol dm^{-3}) .... – 80 cm^3
10 % potassium iodide solution .... – 80 cm^3

Dilute sulphuric acid, starch solution as indicator.

Safety notes

See page 11.

Curriculum links

Redox reactions. Industrial technology. Chlorine.

Possible approaches

To find the amount of chlorine in each bleach:-
The bleach solutions are mixed with potassium iodide and then acidified. The chlorine present in the bleach oxidises iodide to iodine (*ie* chlorine is above iodine in the reactivity series):-

$$Cl_2 + 2I^- \longrightarrow 2Cl^- + I_2$$

The iodine produced is then reduced quantitatively by sodium thiosulphate

$$2S_2O_3{}^{2-} + I_2 \longrightarrow S_4O_6{}^{2-} + 2I^-$$

Thus by adding iodide ions to the bleach and titrating it against sodium thiosulphate it is possible to determine the amount of available chlorine present in the bleach.

The molar ratio $Cl_2 : I_2 : S_2O_3{}^{2-} = 1 : 1 : 2$

TITRATION: 25 cm^3 of bleach in conical flask. Add **(i)** 10 cm^3 of KI solution, **(ii)** 20 cm^3 dilute sulphuric acid (both in excess). Titrate with sodium thiosulphate solution (0.1 mol dm^{-3}) using starch as the indicator (added near the endpoint which is colourless). This procedure uses approximately 10 cm^3 of sodium thiosulphate – the accuracy of the burette readings can be increased by using more dilute solutions of sodium thiosulphate.

You should not give students cost/size of bottle (see data table) – unless asked for. Experiment could be used initially as an assessed practical (*ie* for design purposes where students have to suggest own methods). A couple of lessons later students could carry out their experiment and then present their results for assessment. Credit could be given for accuracy, consistency and treatment of results.

Evaluation of solution

This experiment has been used as an A-level assessed practical (planning stage). Suggest credit is given for:-

1 Clarity of instructions.

2 Identifying a relevant reaction.

3 Choosing appropriate reagents for the reaction.

4 Suggesting appropriate quantities/concentrations of reagents.

5 Selecting appropriate apparatus.

6 Awareness of any variables that may need controlling.

7 Suggesting a way of controlling these variables.

8 Anticipating any hazards that may be encountered.

9 Suggesting ways of avoiding hazards by taking precautions.

After students carry out their experiment, credit could be given for example, for accuracy, consistency and treatment of results under the following headings: Method and explanation. Each titration. Calculation. 'Best buy' value.

Exercise enables considerable discussion regarding consumer choice and advertising.

Suggested write-up

Students write a report for the "Consumer Watch-dog" magazine, explaining their methods and results, and finally which bleach is the best buy!

Data table

Bleach	Cost per bottle	Size of bottle
eg Domestos	65 p	739 cm^3
Vortex		
Parazone		
Sainsbury's		

Extension work

Test viscosity ('staying power'), test 'bleaching power' on, *eg* ink. Discuss/investigate/report on environmental problems due to bleaches. Compare the bleaching power of chlorine and non-chlorine containing bleaches (**NB** Environmental friendly bleaches contain hydrogen peroxide as the oxidising agent that still liberates iodine).

THE ROYAL
SOCIETY OF
CHEMISTRY

Below are germs of ideas for chemical egg races. They have not been fully worked up or tested, but the activities have been included in order to stimulate ideas for your own egg races.

Further ideas for chemical egg races

☞ 'Chemical Magic'. Using all their chemical knowledge students demonstrate a piece of magic. The winner is the one who gets the biggest round of applause!

☞ 'Droopy Bizzy Lizzy' – make a device which shows when a plant needs watering.

☞ Make a (safe!) transportable hand-warmer.

☞ Make a chemical thermometer.

☞ Chemical cooking. Make a scone that rises the most.

☞ Forensic chemistry is a possible avenue to explore:
Who kidnapped Gerald Gerbil?
– ransom note ⟶ chromatography, fluorescent brightening agents on paper
– dust at site of crime ⟶ flame tests
– eye colour in tinted sunglasses ⟶ colour theory

Students could present evidence in 'court' to support/disprove a crime. This could be linked to other school science areas, *eg* Geographical locations – map reading and interpretation, physics – pressure marks on floors.

☞ Find a CHEAP RELIABLE method for cleaning 1p or 2p coins.

Provide: eye protection. Measuring cylinders, plastic droppers, spatulas, stirring rods, various beakers. Balance?

Vinegar (ethanoic acid 1 mol dm^{-3} is fine), sodium hydrogencarbonate, sodium chloride. (Give costs for each reagent.)

☞ Students set up a connected sequence of chemical and physical events (reactions/physical changes) to constitute a "chain reaction" that will run uninterrupted once set off. This is a problem that is open-ended 'with no correct solution', *eg* balls roll down slope, which knocks over test tube so acid dissolves metal, so something falls over, breaking circuit ...

Provide
'junk' apparatus – means to produce gas, blow up balloons, dissolvable materials used as supports, means to produce temperature changes.

☞ Make a Rayon filament capable of lifting an egg.

☞ 10 cm^3 of seawater + simple apparatus to produce maximum distilled water and maximum sodium chloride.

☞ Extracting copper from an ore

Provide
Copper carbonate/sand, rock mixture. Solutions possibly to dissolve out copper carbonate – acids and alkalis. Iron wool or nails.

Objective: Maximum amount of copper from given amount of ore.

☞ What 'salt' or other solid is 'best' to put on the roads to reduce icing? Provide: calcium chloride, sodium chloride, sugar.

☞ My Gran says that Brillo pads won't rust if they are wrapped in Bacofoil. My Grandad says it's because the foil keeps the air out. Investigate whether one or both of them are right.

Provide
Brillo pads with soap, steel wool without soap, Bacofoil, clingfilm.

Your ideas

If you have ideas for chemical egg races then please send them to:

Chemical Egg Races
The Education Department
The Royal Society of Chemistry
Burlington House
Piccadilly
London W1V 0BN

Please submit your ideas on A4 using the headings: age, time, group size, equipment and materials, safety notes, curriculum links, and evalution of the solution. Don't forget to include your name and address.

THE ROYAL
SOCIETY OF
CHEMISTRY

This is to certify that

....

has successfully competed in a
Royal Society of Chemistry
Great Egg Race Competition
held at

....

....

....

on

....